A SYMBOL FOR ALL AMERICANS

On the fifth day, February 23, the Marines took Mount Suribachi. Forty men from the 2d Battalion, 28th Marines, crawled up to the rim of the mountain crater and raised the American flag. Meanwhile, someone hunted around and found a much larger flag and used it for a second flag-raising ceremony. In this ceremony, six men raised the giant flag and they were photographed by Associated Press photographer Joe Rosenthal. This picture first saw light in the photographic dark room on the island of Guam and was to become a symbol for all Americans. The Pulitzer Prize-winning photograph showed five Marines and a Navy corpsman raising the flag. Three of those Marines later died on Iwo Jima, not knowing that they had become very visible in a historic moment.

Bantam Books in The Fighting Elite series

U.S. MARINES
U.S. RANGERS
U.S NAVY SEALS (Forthcoming)

QUANTITY PURCHASES

Companies, professional groups, churches, clubs, and other organizations may qualify for special terms when ordering 24 or more copies of this title. For information, contact the Special Sales Department, Bantam Books, 666 Fifth Avenue, New York, NY 10103. Phone (800) 223-6834. New York State residents call (212) 765-6500.

The Fighting Elite™
U.S. MARINES

Ian Padden

BANTAM BOOKS
TORONTO • NEW YORK • LONDON • SYDNEY • AUCKLAND

THE FIGHTING ELITE®: U.S. MARINES
A Bantam Book / April 1985

*Produced by Bruck Communications, Inc.
157 West 57th Street, New York, NY 10019.*

Cover photo courtesy U.S. Marines

Inside photos courtesy D.A.V.A.

*All rights reserved.
Copyright © 1985 by Bruck Communications, Inc.
This book may not be reproduced in whole or in part, by
mimeograph or any other means, without permission.
For information address: Bantam Books, Inc.*

ISBN 0-553-24702-6

Published simultaneously in the United States and Canada

Bantam Books are published by Bantam Books, Inc. Its trademark, consisting of the words "Bantam Books" and the portrayal of a rooster, is Registered in U.S. Patent and Trademark Office and in other countries. Marca Registrada. Bantam Books, Inc., 666 Fifth Avenue, New York, New York 10103.

PRINTED IN THE UNITED STATES OF AMERICA

H   0 9 8 7 6 5 4 3 2 1

*To all the men and women of the
United States armed forces who
served, suffered, or sacrificed their
lives to keep "Old Glory" free.*

And to Jenny, a warrior's best friend.

Acknowledgments

Many thanks to all the members of the United States Marine Corps, and to the Defense Audio Visual Department, Washington, D.C., who gave their time and provided the historical records and information that made this book possible.

Special thanks must go to the following members of the United States Marine Corps:

Major Fred Peck and Gunnery Sgt. Thomas Griggs, Los Angeles Public Affairs Office, for their guidance and all the historical information they made available to me.

Chief Warrant Officer Ronald Frazier, 29 Palms Public Affairs Office, for taking the time to assist me, by organizing my numerous "visits."

Gunnery Sgt. Marty Richter, 29 Palms, who stuck his neck out for me because of a tank.

Sgt. Ed Burns, Jr., (now happily settled in Iwakuni, Japan), my intrepid desert companion and the "best blooming driver" I ever had.

Sgt. Charles Luedke, another desert rat who never failed to amaze me.

Finally, to all the staff of the Public Affairs Office at 29 Palms, again, many thanks.

IAN PADDEN

Contents

1. Battlefield Log:
   Dong Ha, Vietnam—July, 1966 1

2. History of the Marines 10

3. Battlefield Log:
   The Tartar Wall, Peking, China—May, 1900 17

4. Battlefield Log:
   Belleau Wood, France—June, 1918 27

5. Marine Corps Training 37

6. Marine Infantry
   Small Arms and Squad Deployment 55

7. Battlefield Log:
   Bougainville—November, 1943 65

8. Battlefield Log:
   Iwo Jima—February, 1945 73

9. Mission and Present Structure of the Marines 82

10. Battlefield Log:
    Obong-ni Ridge, Korea—August, 1950 86

11. Battlefield Log:
    Nui Vu, Vietnam—June, 1966 95

12. Women in the Marines 106

| 13 | Future Development of the Marine Corps | 116 |
| 14 | Battlefield Log:
Grenada—October, 1983 | 121 |

1
BATTLEFIELD LOG:
Dong Ha, Vietnam—July, 1966

The barrels of the six 105-mm howitzers were pointing almost straight up. The fire mission had been called:

"Battery . . . adjust.

"Action rear.

"Shell . . . white phosphorus.

"Charge . . . six.

"Fuse . . . quick.

"Center . . . one round.

"Battery . . . one round.

"Shell . . . H.E. [High Explosive]

"High angle.

"In effect.

"Deflection . . . two-four-five-seven.

"Quadrant . . . one-one-three-zero."

The clipped and precise calls came in artillery jargon—always shouted, always repeated, so that there was little room for misunderstanding. The number four gun of the six-gun battery was to fire the first round, white phosphorus, known as "Willie Peter." This was the adjusting round. It would leave a cloud of white smoke after impact and the fire control officer would wait until he was informed that it was on target before he released the high explosive charges which were now snuggled in the breeches of the waiting 105s.

The private who had loaded the Willie Peter glanced at

the skyward-pointing barrel and spoke to the lance corporal holding the firing lanyard. "What are we shooting at? High flyin' groundhogs?"

The lance corporal did not have time to answer.

"Fire!"

The lanyard jerked tight and the gun responded with a fearsome bark. As the recoil mechanism of the gun absorbed the shock of the explosive charge, the thirty-three pound shell was well on its way skyward. The low sound of an incomplete whistle was left in the gunner's ears as the projectile sucked air in behind its upward flight.

As it went skyward it started to slow down, until finally it peaked at a great height, arched gently over, and started its plummet toward its unknown target. Forty-eight seconds after the lanyard had been pulled the shell slammed into the ground at terminal velocity and exploded. It was now some five miles away, and on the opposite side of a mountain, from the gun that had fired it.

The cloud of white smoke that rose from the explosion was watched by the Marine sergeant who had called for the fire mission. He was the leader of a four-man unit of the 1st Force Reconnaissance Company that was operating from a task force headquarters at Dong Ha, Vietnam. Dong Ha was some thirteen miles south of the Demilitarized Zone which divided the north and south of that troubled nation.

The sergeant was lying between a few scraggy bushes in short rippling grass on top of a small ridge. One hundred and fifty feet below him, and some two hundred yards due east of his position, was a small river over which the enemy had built a precarious-looking bridge. The banks of the river on both sides were quite heavily wooded, but the area around the bridge was reasonably clear as the river had chosen to make an S bend at that point and it had created a sandbar.

It was to this area that the sergeant was directing the artillery—specifically to the woods on either side of the bridge. On the previous evening his patrol had seen the enemy crossing backward and forward over the small bridge between the woods. During the night they had heard the voices and sounds of what appeared to be an encamped group of about fifty Viet Cong.

An attempt at a fire mission had been made when they first located the enemy, but the sergeant had called it off as the shells had landed too far away from the target. There had been a strong wind and, although adjustments had been made, it was impossible to maintain the target point—the high trajectory of the shells made their accuracy completely dependent on the will of the wind. The sergeant had decided to wait overnight.

By dawn the wind had abated and, as the light improved, the fire mission was called. The sergeant had asked for one round to adjust, and the Willie Peter had come hurtling in over his patrol with a sound like ripping cloth. It exploded sharply, and as he watched the white smoke the sergeant smiled—it had landed just a little to the right of the position he had called.

"You're gonna get artillery for breakfast, guys," he whispered as he turned to his radio operator. "Have them come left one hundred and fire one volley for effect."

One minute and fifteen seconds later, six high explosive rounds came ripping in and the woods shook with the successive explosions. About fifteen North Vietnamese emerged from the trees on the east bank and waded rapidly across the stream.

"Hit them with a couple of more volleys," whispered the sergeant.

The radio operator quickly transmitted the call and twelve more rounds smashed into the trees on the east bank. The sound of the explosions seemed to mingle in one continuous roar and about forty of the enemy ran from the trees. They splashed and scrambled across the shallow river in apparent disarray.

"Keep that fire coming! Let me have area fire and get them to hit the other bank. Tell them we have discovered a base camp of a mixed VC/North Vietnamese battalion," whispered the sergeant.

By now the enemy seemed to be pouring across the stream away from the east bank. The sergeant turned to his radio operator again and whispered quickly, "Tell them we can now see more than two hundred of them. Have them really pour it on!"

The radio operator's excitement was picked up by the fire control officer and the guns worked in awesome unison as they belched out their cargoes of high explosives.

The gunners toiled and sweated as they maintained a continuous barrage. Small corrections were continuously made to ensure that the shells did not land one on top of another as the area saturation continued.

The sergeant and his three men watched the bombardment. The guns were concentrating their fire in an area only three hundred yards square, the edge of which was only two hundred yards from the patrol's concealed position! The effect was devastating.

As the shells slammed down in random patterns on both sides of the river they felled both trees and North Vietnamese alike. Some of the enemy emerged from the trees onto the river bank carrying packs bulging with supplies, others had their arms full of weapons. At first there seemed to be some sort of organized withdrawal from the trees, but as the artillery pounded the area relentlessly, confusion and panic set in. Some of the enemy just dropped their heavy packs and ran, some ran round in circles.

The patrol watched in amazement as one of the enemy obviously decided that where one shell had landed another was unlikely to fall. He jumped into a shell crater just seconds after it had opened up and huddled down. Another enemy soldier, who appeared to be some kind of officer, ran over to the edge of the crater and immediately started yelling and screaming at the man. It was at that moment that another shell found the crater and enlarged it somewhat, killing both its occupant and his tormentor.

One or two stray rounds landed uncomfortably close to the Marines—it was one of the hazards of calling a close support fire mission.

After forty minutes of the well-directed bombardment, the enemy soldiers were in almost total disarray, most of them panicking. However, some Vietnamese leader, who had decided that the bombardment was too accurate and too consistent to be just random harassing fire, had organized a group of well-disciplined men with the sole purpose of finding out who was directing the artillery.

The sergeant saw about thirty heavily armed men come splashing across the shallow river bed from the west bank. They disappeared into the trees and he could hear them crashing through the thicket up the slope toward his position. It was time to leave—he could now see some of the enemy just 150 feet below him. He quickly informed the guns he could no longer direct the fire and that his patrol could confirm at least fifty-three enemy dead. He would try to get back later, if he managed to evade the searching enemy, and call another fire mission.

The artillery went on standby, the gunners elated. They now knew they had been successful in knocking out some of the enemy and their morale had improved.

The patrol left the top of the hill at a fast pace and made no attempt to conceal its back trail—there was no time. It came down the back of the hill away from the enemy camp in a southwesterly direction. One of the corporals found a small stream bed and the soldiers were grateful for its protecting shadows as they raced away from their pursuers. Running downhill jarred every bone in their bodies, particularly their knees, which threatened to collapse at any moment.

It was only their excellent physical condition and experience at working over rough terrain that allowed them to maintain such a fast pace. Despite the fact that their equipment was well strapped to their bodies, everything seemed to move. The water-laden canteens, the grenades, and the ammunition pouches were the worst—they bounced cruelly against their bodies but did no damage as the soldiers' bodies were protected by hard callous pads developed from months of patrolling in the harsh terrain. The sweat ran in rivulets over their greased, camouflaged faces and the salty liquid trickled into their eyes, ears, and mouths.

At the task force headquarters, the patrol's operations officer had been listening to the whispered radio conversations and had heard the situation develop. He acted immediately to provide his patrol with support.

The pilot of a small observation plane in the vicinity was asked to fly up the valley toward the patrol's position and instructed to provide whatever assistance he could. A Marine jet—an F-8 Crusader—with a two-thousand pound bomb under each wing was directed to the scene and ordered to take instructions from the observation plane.

By this time the patrol had covered almost a thousand yards and was now at the bottom of a hill where the men collapsed, almost completely winded. They were extremely uncertain about their future as they lay gasping to regain their breath. The enemy, they reasoned, was close—not more than four or five hundred yards behind them—and they would have no difficulty in picking up their trail. Indeed, even someone who had lived his life in a big city could have instantly picked up the trail in the dry grass on the hillside. The tracks in the stream bed were not so clear, but they would present no difficulty to the experienced enemy.

The pilot of the observation plane was a skilled veteran. He spotted the base camp that had been devastated by the guns; he counted fifty bodies. He flew low, no more than fifty feet above the trees, around the hill and toward the estimated position of the recon patrol.

The sergeant heard the soft, steady sound of the single-engined observation plane and stood up. Seconds later he saw it coming straight toward their position. He waved and flashed his signal mirror at the plane and received an immediate response—a quick rocking motion of the wings from the tiny plane.

The pilot contacted the patrol on its radio frequency and the sergeant explained his plight. "They're right behind us . . . on the hill!"

The pilot acknowledged and swung toward the hill. His quick eyes glimpsed the North Vietnamese as he passed low across the hill. The sergeant was right—the enemy was close, too close!

He radioed the sergeant and told him to take cover quickly. He called the F-8 as he turned toward the hill again and fired two marking rockets right in front of the pursuing enemy.

The F-8 pilot saw the thin white smoke wisps of the

observation pilot's marking rockets. His target was quite clear, but he became very uneasy when the observation pilot told him that the marines were only three hundred yards away from that target. He listened intently to the observation pilot's instructions. "Hit fifty meters at twelve o'clock on my smoke on a holding of two-one-zero degrees."

As he eased his thundering jet into a thirty-degree dive, the Marine pilot remembered that he had once dropped one-thousand pound bombs within seven hundred yards of his own troops. But two-thousand pounders at three hundred yards? He hoped the patrol were keeping their heads down.

As his speed increased in the dive, he was thankful that the Crusader was a steady-aiming aircraft. At over three hundred miles an hour he approached his drop point, and at eighteen hundred feet above the ground he pulled the aircraft very sharply and smartly out of its dive. He felt the load on his body as his accelerometer registered five Gs. Both of the two-thousand pounders—thick, cylindrical, wicked-looking hulks painted a drab green—fell free from beneath the wings and plummeted down.

The sergeant and his men were listening intently for the sound of the enemy when the sharp screech of the F-8 filled the air—they never even saw the howling jet as their instant response was to seek further shelter, shelter which was to prove only marginally safe.

The pilot of the observation plane watched the Crusader on his bomb run and smiled with satisfaction when he saw the veritable blockbusters fall from beneath the wings. The bombs wobbled a little at first, then stabilized as they headed toward the target markers.

The enemy, who knew they were close to the recon patrol, both heard and saw the Crusader. The pilot of the observation plane watched their faces as they looked skyward at the falling bombs. He shook his head in wonder. He had looked down upon similar scenes many times before and it always amazed him how troops looked up and appeared to be mesmerized by falling munitions. Them again, he reasoned, why shouldn't they? They were watching their own destruction.

The two fearsome weapons were on target and they exploded on impact. The ground jumped beneath the marines. They were severely jolted by the massive blast and their already-strained lungs were further tortured in the shock wave. They stared, open mouthed, at each other; they could hear nothing as a high-ringing noise filled their ears. For a moment they were not sure what had happened, then dirt, boulders, limbs of trees, and almost complete tree trunks started to fall around them. They scrambled for further cover, knowing that most of the debris could kill them as easily as any bullet.

When the torrent of foliage and earth had ceased, the radio operator whispered furiously to the observation pilot. "You damn well missed, you almost hit us. Hit further back, dammit! Don't hit here again!"

The calm voice of the observation plane pilot came back over the radio. "Sorry, guys. Had no choice. Your friend was right on target. Nobody's behind you now. Get going while you got a lead on."

The pilot was right, the pursuing enemy had been completely obliterated by the well-placed bombs. Ravaged remains were testimony to that.

The sergeant had no intention of pulling out completely. He attempted for several hours to get his patrol back into another observation position over the North Vietnamese base camp. But it was not to be—the enemy had sent further patrols out to search for them.

The Marines turned east to evade the enemy, who could now be heard searching for their tracks. After a few hours the sergeant determined that the area behind them was well covered by enemy patrols who were slowly getting closer. He called for help from the artillery and the first Willie Peter came ripping in not one minute after his call.

The gunners were excited again, they knew the call for the fire mission was from the same patrol they had successfully assisted earlier in the day. The barrels were pointing skyward again, and the gun crews waited eagerly as the Willie Peter took its forty-eight seconds to reach the target area.

When the corrections were called, the crews realized

instantly that Willie had found the mark. The lanyards jerked and the six howitzers recoiled almost as one. The breeches opened up like the mouths of baby vultures waiting to be fed. And fed they were, quickly and efficiently, round upon round.

For nearly an hour the crews maintained their barrage and the recon patrol watched volley after volley smash into the area. Between successive volleys they could hear the cries of the enemy, who were finally persuaded that it really wasn't worth it to pursue the marines any further.

After the bombardment, the patrol neither heard nor saw the enemy again. They headed northeast for the remainder of the afternoon and as night fell they found a "harbor" location in the dense thicket on the side of a hill. They were almost at the exhaustion point and had great difficulty staying awake on watch throughout the night. Moving around was out of the question—noise travels far at night, particularly in brush and jungle. The sergeant, through sheer determination, stayed awake for most of the night to ensure that those on watch did not fall asleep.

Dawn found them on the move again, this time in a southwesterly direction. They were very tired but in high spirits as they were now moving toward an "extraction point." Their mission was completed and a helicopter was on its way to pick them up.

At eight o'clock that morning, July 29, 1966, they guided the helicopter into a safe area and were quickly lifted out to their task force headquarters. The four-man reconnaissance patrol had just called one of the most successful artillery missions on record in the Vietnam War.

2
HISTORY OF THE MARINES

The United States is a maritime nation with worldwide interests. It has been so since its birth and it is not likely to change any time soon. As a result, the United States Marine Corps has become the nation's amphibious force in readiness.

The word *marine* originates from the Latin word *marinus*, meaning "sea." The modern word *amphibious* means "living, existing, or operating on both land and in the sea," and it comes from the Greek word *amphibian*, which means "living a double life." What follows will show that both of these words are appropriately used in describing the United States Marine Corps.

The U.S. government, in the National Security Act of 1947, decreed that

> The United States Marine Corps, within the Department of the Navy, shall include land combat and service forces and such aviation as may be organic therein. The Marine Corps shall be organized, trained, and equipped to provide fleet marine forces of combined arms, together with supporting air components, for service with the fleet in the seizure or defense of advanced naval bases and for the conduct of such land operations as may be essential to the prosecution of a naval campaign. . . . In addition, the Marine Corps

shall provide detachments and organizations for service on armed vessels of the Navy, shall provide security detachments for the protection of naval property at naval stations and bases, and shall perform such other duties as the President may direct: *Provided* That such additional duties shall not detract from or interfere with the operations for which the Marine Corps is primarily organized. The Marine Corps shall be responsible, in accordance with integrated joint mobilization plans, for the expansion of peace-time components of the Marine Corps to meet the needs of war.

One does not have to be an expert in legal or bureaucratic jargon to understand this section of the National Security Act. The mission of the Marines is quite clear.

It is interesting to note that, despite the fact that the Marine Corps is an integral part of the navy, an amendment was made to the National Security Act in 1978 giving the commandant of the corps the status of "full member" of the Joint Chiefs of Staff. It was an indication of the importance, respect, and trust that the United States government places in the Marine Corps.

Why was the Corps given such privileged treatment? There is no one incident or action that earned them the status they were afforded, and still have. However, if we take a look at the history of the Corps the reasons will become obvious.

At least five centuries before the Christian era, the Phoenicians and all the Greek maritime states carried complements of soldiers, in addition to their normal seafaring personnel, on board their ships of war. The task of those soldiers was to fight in naval engagements defending the sides of their own ships when they were locked in battle with an enemy vessel. They were also used as the boarding parties during attacks on enemy vessels and, at times, they were put ashore to hold the land approaches to harbors when their fleets were using the ports of strange countries.

They were further used as raiding parties on land when the fleet commanders wished to strike enemy towns or settlements.

The practice of using soldiers on board ships of war seems to have passed down from the Greeks and Romans to the British, who used them extensively. But it was not until the reign of Charles II that a military organization was placed under the direct authority of the British Admiralty. On the advice of the Duke of York and Albany, who was at the time Lord High Admiral of England, Charles II, on October 28, 1664, issued an "order in council," directing the organization of "The Admiral's Maritime Regiment." Shortly afterward it was renamed "The Regiment of Marines," becoming the first troops to join the British fleet vessels for service at sea in times of war.

During the next seventy-six years, the British Marines were never formally organized to any extent, although each man-of-war did have its individual complement. In 1740, the British Marines were formally re-established by King George II; at the same time the American Colonial Marines were founded. Three regiments were raised in America, in conjunction with the British regiments, for service in the West Indies under Admiral Edward Vernon of the Royal Navy. It was felt that native American Marines were better suited to the West Indian climate than their European counterparts, and that their uniforms of "camlet coats, brown linen waistcoats, and canvas trousers" were ideal for their particular duties.

On April 2, 1740, Alexander Spotswood, former governor of the colony of Virginia, was commissioned by King George II as a major-general and became the commandant of the entire American Marine organization. Colonel Spotswood died in June 1740, and the entire Marine force was then formed into a single regiment with four battalions. Command was then given to Colonel William Gooch of Virginia and the regiment picked up the name "Gooch's Marines." It was officially known by the British army as their 43rd Regiment. (One of the officers of the regiment was Lawrence Washington, the half brother of George.)

The joint service of the British and American units in

the Caribbean during the 1740s is recognized to this day by both services in their use of the colors gold and scarlet.

November 10 is recognized as the official birthday of the Marines because it was on that day in 1775 that the Continental Congress passed an act making the Marines a regular branch of the nation's armed forces.

During the American Revolution, there were few Marines, but they gave valued and notable service. Their exploits ranged from serving with the naval forces on expeditions to the Bahamas, to serving with John Paul Jones, and with Washington's army in the Battle of Princeton and Assunpink Creek. At the end of the Revolutionary War, both the Marines and the Navy were disbanded.

It was not until the naval war with France in 1798 that the Marine Corps, as it exists today, was formed. The corps was authorized by Secretary of War James McHenry, who also authorized the now-famous blue dress-uniform of the corps.

Although it is not usually associated with the uniform, the term *Leathernecks*, which is the nickname for the marines, probably comes from the black leather collar, or stock, which was supposedly worn between 1798 and 1880 by the early marines. Although it has never been substantiated, it is said that the collar served to protect the jugular vein from a sword slash. However, the British Royal Marines did have the same nickname, because they wore a similar collar. More recently the sailors of the Royal Navy have changed the name to "Bootneck," and occasionally shorten it further to "Bootie."

It was 1801 before the American Marines finished the Quasi-War with France. In that same year, they became involved in the fight with the Barbary corsairs, in and around Tripoli on the North African coast. This skirmish lasted until 1815.

In the War of 1812, the Marines served on almost every American warship that engaged the enemy. They also served with the army in Maryland in the Battle of Bladensburg in 1814, and with General Jackson at the Battle of New Orleans.

Raids against pirate strongholds in Cuba were carried

out in 1824, and in 1832, after Malay pirates had plundered the USS *Friendship*, the Marines landed at Kuala Batu on the island of Sumatra as part of a joint force from the U.S. warship *Potomac*.

A mutiny in Massachusetts State Prison in 1833 was beyond the control of civil authorities and Marines from the Boston Navy Yard were called in to suppress it.

In 1836 and 1837, the Army was fighting the Creek and Seminole Indians in Florida and Georgia. They were assisted by the Marines who served under their own Commandant, Colonel Archibald Henderson.

The Marines saw extensive action during the wars with Mexico and in the conquest of California. They were in action on both the Atlantic and Pacific coasts and were involved in the capture of San Francisco, Monterey, Tampico, Vera Cruz, Mazatlan, and Tabasco.

When Major General Scott marched into Mexico City, a battalion of Marines accompanied him. They took part in the final attack on Chapultepec Castle and the taking of the Mexican National Palace, the "Halls of Montezuma."

For a number of years after the war with Mexico, the Marine Corps's flags carried the words, "From Tripoli to the Halls of the Montezumas." This was the so-called Tripoli–Montezuma flag and it carried the motto, "By Land, By Sea." In 1868, the present-day Marine Corps emblem was adopted and the Navy Department authorized the use of the words on the Marine Corps's flags. Shortly after the Civil War, the motto on the flags was changed to *Semper Fidelis* (Latin for "Always Faithful"), and it was officially adopted as the motto of the Corps in 1883.

The first two lines of "The Marine's Hymn" commemorate the campaigns in North Africa and Mexico. The unknown author of the hymn reversed the order of events—Tripoli actually came before Montezuma, but the song just wouldn't sound quite the same if it were sung that way. In 1888, John Philip Sousa, who was the leader of the Marine Band, composed the marines' march, "Semper Fidelis."

When the Civil War ravaged the nation, the Marines saw distinguished service both ashore and afloat. After the war they were used to quell the labor riots in Baltimore and

Philadelphia, and assisted with the troublesome enforcement of the revenue laws in New York.

Between 1867 and 1882 they saw global service, protecting Americans and American interests, most notably in the Caribbean, Korea, China, and Egypt. It was shortly after this period that the war with Spain began and the Marines were in action on the ships of Admiral Dewey and Admiral Sampson in the Battle of Manila Bay and the Battle of Santiago de Cuba. After the victory in Manila Bay, First Lieutenant Dion Williams led the Marines of the battle cruiser *Baltimore* in the capture of the Spanish naval arsenal at Cavite.

Shortly after the war with Spain, the Marines were sent to assist in suppressing the insurrection in the Philippines and, in 1900, during the Boxer Rebellion in China, they assisted in the defense of the international Legation Quarter in Peking. During that same rebellion, they fought alongside the British Royal Welsh Fusiliers in the battle for Tientsin. On various occasions each of these fighting elite assisted one another and a bond of admiration and respect developed between them. That bond has continued to this day and is recognized on the first of March every year by the exchange of cabled greetings that carries only the old Welsh password, "And Saint David." (Saint David is the Patron Saint of Wales, and March 1 is celebrated by Welshmen throughout the world as Saint David's Day.)

Between 1903 and 1914 the Marines were in action in Abyssinia, Santo Domingo, Korea, Cuba, Nicaragua, Panama, and Vera Cruz, Mexico.

The First World War took the Marines into the heart of Europe as an infantry brigade serving with the Second American Division. At the Battle of Belleau Wood in June 1918, the 4th Brigade was involved in a ferocious battle with the Germans. The Marines fought with such tenacity that the German official reports described them as *"Teufelhunden,"* or "Devil-dogs." That name is still being used.

Between the First and Second world wars, the Marines saw continuous action in the Caribbean—Haiti, the Dominican Republic, and Nicaragua being the principal areas of engagement.

China became more troublesome again in that same period and the Marines saw continuous action there until they were withdrawn in December 1941.

"Climb Mount Nikita" was the coded message sent to Admiral Yamamoto to commence the attack on Pearl Harbor. When the devastating attack on the unsuspecting Americans was over, Yamamoto discovered that his attack had preceded the Japanese Declaration of War. Upon hearing this, he is reported to have said, "I am afraid that we have awakened a sleeping giant who will be filled with a great resolve." He was right—and it was the Marines who became the spearhead, the boot, and the fist of that giant, as it slowly and painfully battered the enemy back to the shores of Nippon.

From Wake Island, through Guadalcanal, the Solomons, the Gilberts, the Marshalls, the Marianas, Palau Islands, the Philippines, and Okinawa, the Marines added to their legend.

Their legend continued to grow through the Second World War, the Korean War, the bloody and ignominious Vietnam War, Lebanon, and the island of Grenada.

This brief history shows that the Marines have actually been fighting for the nation since before its inception. They have continued to earn respect since those early days, and they have provided such valuable military service to the nation that it was only natural that the Congress of the United States should recognize it and award the commandant of the Corps an equal say on the highest military council in the land, the Joint Chiefs of Staff.

3
BATTLEFIELD LOG:
The Tartar Wall, Peking, China—May, 1900

"Fighting packs, rifles, and bayonets . . . and they'd better be sharp," bellowed Sergeant Joe MacClintock.

The group of Marines receiving the orders were on board the USS *Oregon*, lying at anchor some twelve miles off Taku bar on China's northeast coast.

The young Marines were silent for a moment. They had lain idly at anchor for days, impatient to get ashore. The order had now come and it was plain from their veteran sergeant's remark that action was expected.

A young, eager Marine asked the obvious, "Think we'll see some action, sergeant?"

MacClintock looked at him gravely and thought he saw a shiver of anticipation in the youngster. "Sooner or later, we will. We're Marines and that's why we're here."

MacClintock turned away as the Marines readied their packs. He was a little concerned about the men in his platoon. There were too many youngsters, recruits, immature boys with romantic notions about combat and life in the Marine Corps. He wondered how they would stand up under the pressure of an enemy attack, and he wished that a few more salty old veterans had been assigned to the platoon.

He brushed aside his doubts as he looked over the

blue-gray waters to the coast of China. The fanatical arm of the Chinese Imperial Army, "The Fists of Harmonious Righteousness"—the Boxers—were in revolt. They were pledged to the destruction of all "yang kuei-tzu"—the "foreign devils."

It was too late to worry now. These young Marines were on their way to Peking. He would just have to train them further in what little time was left. Perhaps the Marine Corps's fighting spirit would come through? He hoped so, because he knew from experience that when the bullets started slamming around them, each man would be on his own.

MacClintock and his Marines from the *Oregon* joined Marines from the cruiser *Newark*. The fifty-strong company was to be led by Captain John Myers. The Marines were joined by five sailors who were to man the Colt machine gun they had brought with them. Captain Myers ordered that combat packs only would be carried, no baggage. Instead, the small group carried between them some twenty thousand rounds of rifle ammunition and eight thousand rounds for the Colt. In the weeks that were to follow, this was to prove to be a very astute decision.

The Marines were part of an allied contingent that was being sent to Peking to protect the lives of the foreign diplomats and their families who were stationed in the capital.

At eight o'clock on the evening of May 31, 1900, 337 officers and men of the allied nations arrived by train from Tientsin and marched through the streets of Peking with fixed bayonets. Leading the group were the fifty United States Marines, and Sergeant MacClintock could be heard calling cadence to his men, "You're supposed to be Marines. Look the part!"

In 1900, Peking was literally a collection of walled cities within a city. Within its bounds were the Tartar City and the Chinese City. In the center of the Tartar City was the Imperial City and inside that was the Forbidden City, where Sacred Lions protected the buildings.

Around the Tartar City was a massive wall, some forty feet high and forty feet wide at the base. The southern wall of the city was also the northern wall of the Chinese City.

In the center of the inner sanctuary of this collection of walls (inside the Imperial City) stood the palace of the rulers of China, the descendants of the Manchu dynasty. Just outside the walls of the palace were the legation quarters of eleven nations, located on a plot of land approximately fifteen hundred yards square. Through the center of the Legation Quarter, running from north to south, ran a small canal. The southern boundary of the Legation Quarter was bounded by the great Tartar Wall; the canal exited the city through a tunnel and sluice gate in the wall.

The American legation nestled alongside the wall and was bordered on its east side by the canal. Directly opposite the American legation, to the north and slightly west across Legation Street, was the Russian legation. To the north of that was the British legation, the largest of all, and just to the north of that was the southern wall of the Imperial City. On the east side of the canal were the legations of France, Austria, Hungary, Holland, Italy, Belgium, Spain, and Germany.

The containment of all foreigners in this manner was in keeping with the orderliness of the Manchus. The proximity of the Legation Quarter to the palace was also to their liking—foreigners were best kept where they could be watched by the rulers.

In 1898, the ruler of China, the young Emperor Kuang Hsu, was deposed in a bloodless coup by his aunt, the Empress Dowager, Tz'u-hsi. At the request of the old conservative members of the court, she simply walked in and told the "Lord of Ten Thousand Years" that she was taking over. He offered no resistance. The nation's elders and the imperial armies were loyal to her and she was no stranger to the court—this was the third time in her life she assumed the rule of China.

Tz'u-hsi was sixty-four years old, although she looked about twenty years younger, and had come to be known as "Old Buddha." She had made no serious attempts to keep the reactionary factions under control; in fact she courted them, but not openly, as they wished to return to the "old ways," which she also favored.

The first task, and the burning desire of the zealots of

"The Fists of Harmonious Righteousness," was to rid the nation of non-Chinese and the Christian faith.

Within two years of Tz'u-hsi taking charge the strife had hit the diplomats of all nations, particularly those in the city of Peking. It had started with sporadic riots accompanied by the usual rock throwing and window smashing. When the shooting started, the diplomats called for armed support.

The arrival of the soldiers gave the legation community hope. At a joint meeting of the senior diplomats it was decided to negotiate for the safe evacuation of all legation personnel to Tientsin. The Chinese had offered safe passage, but would not guarantee it, and finally issued an ultimatum: all foreigners must leave Peking by five o'clock on June 20. The legation members had decided they would never get to Tientsin alive if they tried. The Boxers would kill them all en route.

The Legation Quarter was prepared for siege. As the deadline was reached, the Old Buddha declared war and ordered the regular elements of the Chinese Imperial Army to join with the Boxers in attacking the legation.

The eight-week siege of the Peking Legation Quarter opened up with a barrage of artillery and rifle fire on the legation defenses.

The Marine's area of defense was around their own legation and on the Tartar Wall overlooking the Chinese city. The battle quickly turned into something of an endurance contest for control of the wall.

Half of the detachment manned the wall itself and the remaining half took up defensive positions on the ground behind. The Chinese mounted only sporadic charges—they preferred to keep the legation under continuous artillery and rifle fire as they burned and destroyed the surrounding area.

The first serious charge on the wall came on June 27. Hordes of pigtailed Boxers, with the traditional crimson sash around their waists and most with similar-colored headbands, stormed out from behind their own barricades toward the Marine position. The accuracy of the Marine fire quickly cut down the first wave of the seemingly crazed enemy. The second wave was treated in a similar manner and, just when it seemed that they were about to breach the defenses en masse, the charge stopped. Captain Myers, Captain Newt Hall, and Sergeant MacClintock were on the wall as the charge started and it was through their encouragement that the Marine line held fast. Marine Privates Sommers and Daly had blood on their bayonets from four of the enemy who had breached the defenses.

The Chinese then started to build a series of barricades which, by July 1, had closed to within a few feet of the wall. They were so close, in fact, that they could hurl large rocks at the marines with a mechanical contraption not unlike the ancient Roman catapults. Then the Boxers started an artillery bombardment of the wall from close range and the Marines were forced to withdraw.

The American minister, Edwin Conger, and the British minister, Sir Claude MacDonald, who had been elected defense commander, ordered Captain Myers to retake the wall. Both ministers were ex-military men and they feared that with command of the wall the Boxers would create havoc with their field pieces and sniper fire.

Myers asked for reinforcements to assist and he was given a detachment of British Royal Marines and some Russians. The wall was easily retaken, but the Chinese still held their barricades close to it.

Edwin Conger and Sir Claude MacDonald conferred again and summoned Captain Myers. Myers couldn't believe the orders he received this time: he was to take fifteen Marines, twenty-six Royal Marines, and fifteen Russian soldiers and drive the Chinese back from the wall. The American Marines were ready. The Royal Marines were quite willing and eager—they trusted Myers's leadership, and said so. The Russians went along with the plan, basically because they had no option.

When Myers asked for some time to work out the details, he was told that the raid must take place that night.

In the early hours of the morning of July 3, Captain Myers and Sergeant MacClintock led the troops over their own barricades toward the Chinese positions. One of the American Marines and a Royal Marine quickly silenced the first two sentries with their razor-sharp knives, and the small party was amongst the Boxers before the Chinese knew what was happening. The surprise was total and the Chinese fled, leaving fifty dead. Two Marines died and several were wounded, including Captain Myers, but the Marines were elated, and so were the people inside the legation. The wall defenses were secure again and they were now widened and strengthened.

Someone discovered an old gun barrel in a disused foundry on the legation. Sergeant MacClintock was called to inspect the rusted old barrel. MacClintock found that the barrel was rifled and that with a little work they might be able to do something with it. It would be good for morale, if nothing else.

The Italians had some spare wheels on which to mount the gun and it was discovered that the Russian nine-pound gun charges would fit the breech perfectly. When all the rust was chipped away and the gun was cleaned and prepared, MacClintock named the weapon "Betsey."

The first firing of Betsey was conducted with some ceremony and doubt, as the gun barrel was at least forty years old. The gun was too inaccurate to be effective at long distances, but it was most effective at close range—particularly as it was charged with old nails, nuts, bolts, pieces of scrap iron, and chain. Such a charge could never be placed in the barrels of one of the better guns, but Betsey's belly could handle it and then spew it out with great effect. The gun caused a great deal of unhappiness amongst the Boxers as it dispensed a massive swath of lethal junk every time it was fired. Betsey did as much to bolster the morale of the defenders as it did to demoralize the attackers.

The first four Marines to die were killed by sniper fire, and throughout the siege snipers continued to take their toll. However, the marksmanship of the Marines had been well demonstrated; it was one of the things most remembered by those who survived the siege.

It was decided that the defenses below the wall to the east of the American legation needed to be built up and strengthened. The builders would need covering fire to keep the enemy snipers away. Private Dan Daly was chosen by Captain Newt Hall, who had taken most of the responsibilities of the wounded Captain Myers, to man the wall on the night of July 12.

As Hall and Daly moved along the wall to locate a good covering position for Daly, they were attacked by a horde of Boxers. The two Marines drove them back and off the wall after killing four of them. Hall was now worried that Daly would be in great danger if left by himself and offered to send up a few more Marines. Daly insisted that he would rather be there alone—he felt it would be easier if he only had to concern himself with the enemy.

Although Daly had only been in the Marine Corps for a year, he had already developed some excellent soldierly qualities, and he was an adept marksman. Hall had been in action alongside him earlier in the siege and had been impressed by his calm professionalism. Hall had also noticed that the crusty Sergeant MacClintock had a good deal of respect for the young Marine. With these thoughts in mind, Captain Hall left Private Daly on the wall by himself.

As Hall moved back to relative safety, Daly settled into his position. Within an hour he had shot eight enemy snipers. Although a few stray shots had been fired in his direction, nothing much troubled him. The Boxers knew where he was and Daly wondered why he had not yet been attacked, but he was convinced that sooner or later the Boxers would come.

For nearly thirty minutes no further targets presented themselves to him, but he heard the enemy scaling the wall—they were coming after him.

Six Boxers charged toward him. Three quick shots stopped three of them. The fourth man took the bayonet on Daly's rifle in the neck. As Daly pulled his bayonet out, a fifth Boxer was on him. He swung the butt of his rifle into the enemy's face and the man spun backward. The sixth attacker took a full thrust of the marine's bayonet. Daly then used his bayonet to finish off the enemy he had clubbed. He carefully checked the bodies of the remainder to make sure they were dead.

Daly settled back down into his position and absentmindedly cleaned his bayonet. He checked his rifle carefully and became a little annoyed—he did not like using the weapon as a club.

Daly's breathing rate had increased after the attack and it had just settled down when he heard a call from Captain Hall asking if he was safe. The attack had been heard by the Marines behind the main defenses and they were concerned for their colleague's safety. His answer was a well-aimed shot at another sniper whose cry of pain, followed by his crash to the ground, told the captain all he wanted to know.

Two more snipers fell to the marksman before he was attacked again. This time, four ugly-looking Chinese managed to get a little closer before Daly spotted them. As they charged, he shot two of them and the third ran into his bayonet. The fourth one hesitated for a moment and was about to continue when Daly shot him. Again, he checked the bodies and cleaned his bayonet. He was surprised that none of the enemy had carried a gun; he was also thankful.

Another call came from Captain Hall and the reply was the same as before—Daly picked off another sniper.

Nearly an hour had passed and the builders worked unhindered.

Daly moved his position slightly and found himself another target—a Boxer moving stealthily between two barricades. The man fell without a sound and lay still.

Daly heard a sound to his right and turned in time to see four men charge. He had time for one quick shot as they were almost upon him. He saw one man fall and then saw the flash and felt the blast of a rifle fired at close range. He felt no pain, but the Boxer who had fired the shot assumed he had hit the marine. For his assumption he took Daly's second bullet in the chest. The third attacker hadn't hesitated and pushed the barrel of the Marine's rifle to one side in preparation for a savage thrust with a wicked-looking knife. As the barrel of his weapon was thrust away, Daly swung the butt up hard into the enemy's throat and a great throttled gasp ensued. The man was sinking to his knees as the remaining attacker approached. He hesitated for a moment, then turned as if to run away. Daly shot him in the side of the head and he slumped to the ground.

Breathing heavily, he put his knife attacker out of his misery and again checked the bodies. He felt something trickle down inside his shirt on his left side and he remembered the shot that had been fired at him. Daly had heard stories about Marines who had been mortally wounded and had not noticed it until a calm period in the fight. As he groped around under his shirt, he became a little scared. He pulled his hand out, it was soaked and he felt the liquid trickle on his fingers. He could not see any real color as his hands were dirty from crawling around the wall. He began to feel weak, but he could still feel no pain. He smelled his hand, then tasted the liquid—sweat! Daly sat down and laughed softly to himself for a few minutes before returning to his favorite position.

The fourth attack came before he fired at another sniper. This time, there were six more crazed Boxers charging at him. They had not managed to get as close and he shot three of them before the others turned and fled.

Ten minutes later, he shot another luckless sniper and was about to take aim on still another when he heard a wild scream close behind him. He turned and fired at the fast-moving target. He missed, but he did not miss with the bayonet. The enemy stopped and hung grotesquely on the bayonet as he gave out with a gurgled scream that sprayed Daly with blood. He pulled the trigger again and blew the man off his rifle. As he did so, another attacker appeared and he, too, received the bayonet.

During the rest of the night, Daly repulsed another three attacks; he killed at least nine more of the enemy and drove off the remainder.

As dawn broke, Daly crawled behind a pile of bodies back to his own barricades. Red-eyed, tired, and covered in stale enemy blood, he dragged his way off the wall. In the light of day, Captain Hall and his fellow Marines saw the carnage.

From that time on, the Marines had no difficulty in holding the Tartar Wall. They were still there when the American and British troops of the relief force reached them.

After a heroic fifty-five day siege, the international

Legation Quarter was finally relieved. The Marines had suffered seven dead and ten wounded, but they had performed remarkably well. They had held up under fire, despite shelling, sniping, sickness, fatigue, and a diet of dirty rice and horse meat.

Private Dan Daly was awarded the Medal of Honor for his night on the wall. He was to win another Medal of Honor in the conflict on Haiti in 1915. But Dan Daly, who was not a talker, will best be remembered amongst Marines for a statement he made in France at the Battle of Belleau Wood during World War I. As a sergeant, he was urging his platoon to move off faster into the attack and he bellowed out, "Come on, you bastards, do you want to live forever?"

4
BATTLEFIELD LOG:
Belleau Wood, France—June, 1918

By the end of 1916, the First World War had gone badly for the combined British, French, and Russian allies.

In the North Atlantic, German submarines had been on the rampage in their attempts to stem the flow of American war supplies to Europe, and they were indulging in a campaign of unrestricted warfare on Allied and neutral shipping.

President Woodrow Wilson and most of Congress were convinced that America would soon get involved in the fighting if the Germans continued with their unrestricted attacks. The decision was hastened when the great British ocean liner *Lusitania* was torpedoed off the Old Head of Kinsale in southern Ireland. On board at the time were a considerable number of American citizens, and the outrage at home was such that the United States became committed and started to prepare its armies.

The war in Europe was taking its toll on the Allies. Revolution in Russia had removed opposition to the German Eastern Front, and the Italian army to the south was being badly mauled. Consequently, the Germans were redeploying their forces on the Western Front, much to the dismay of the embattled British and French armies.

On February 17, 1917, as America was still preparing its armies, British and French authorities visited Washington and begged President Wilson to send at least a token force to boost the sagging Allied morale. At the end of May, 1917, the President agreed to the British and French request and chose the army's competent and reliable Major General John Pershing to lead the American Expeditionary Forces.

The commandant of the Marine Corps, General George Barnett, despite the protests of General Pershing and several members of the Army staff, had somehow managed to persuade the War Department to include a Marine regiment in the first expeditionary force.

Barnett was trying to build the size and strength of the Corps, and had earlier seized the opportunity of the threat of war to achieve this end. He had been granted permission by Congress and the Navy to recruit and enlist 12,500 additional Marines. With the Marines' claim of being "First to Fight" and with his massive recruiting drive underway, Barnett realized that he could lose funding if he did not have the Marines in combat as soon as possible.

Pershing argued logically that he did not need the Marines since this was not an amphibious effort, and he had sufficient Army troops for his initial purpose.

Barnett, however, had gone even further. With the help of Secretary of the Navy Josephus Daniels and Chief of Naval Operations William Benson, he had also persuaded the War Department to accept a further regiment as soon as the first contingent of the expeditionary force settled in France. The Marine complement to the force was now a full infantry brigade of six thousand men organized in the same manner as the Army.

General Pershing was upset since he had no choice in the matter. His own Army chief of staff, Tasker Bliss, and the secretary of war, Newton Baker, made the decision. In return for providing the Marine regiments and necessary replacements, the War Department promised Barnett that they would supply the Marines with any weapons and equipment necessary to bring them up to Army requirements. General Pershing had to promise that he would treat the Marines equally in all supply and personnel matters, and Secretary of War Baker promised that the Marine brigade would be assigned to a combat division.

The Army, however, could not supply the transport ships to get the Marines to France, and the Navy immediately stepped in to transport the Marines on U.S. warships.

While General Pershing was in the final stages of preparing his division for embarkation, the newspapers of the day announced that the Marines would definitely be the first to fight as they had provided the expeditionary force with a regiment of veteran fighters from the Caribbean campaigns. General Pershing wasn't pleased.

In June of 1917, the 5th Regiment of the 4th Brigade of the Marine Corps sailed for France as part of the army's 1st Division. The Marines landed in France eager to begin serious field training in preparation for their move to the Western Front, but they were to be disappointed.

For whatever reasons, the army and General Pershing's staff had greatly underestimated the American Expeditionary Forces' need for support element troops. Pershing therefore assigned the Marines as security detachments and labor troops. He did not do it because of any lack of confidence in the Marines—in fact, he admired their smart appearance and discipline. However, the 5th Regiment was actually surplus to the 1st Division, and Pershing did not have a very urgent requirement for troops in the support element.

The General Staff of the War Department agreed with Pershing. They felt that the Marines would make excellent provost guards, and, as the 1st Division trained for the front line, it was never accompanied by more than a battalion of Marines.

The Marines performed their rear-echelon duties well, but they were upset and annoyed, as was their commandant.

When the 6th Regiment arrived in France, it, too, was assigned duties that were similar to the 5th's. But at least the 4th Brigade was now complete.

Prior to the arrival of the 6th Regiment in France, General Pershing had reassigned the brigade to the Army's 2d Division. The Marines had already lost faith in the Army staff, and they were convinced that they were being deliberately frustrated. This only served to increase the intensity of the Marines' fervor to be outstanding. This fervor was so intense that football games were banned by Brigadier General Charles Doyen, commander of the brigade, because they were being played with a little too much "enthusiasm."

Commandant Barnett requested of General Pershing's staff that the Marines not be swallowed up in name with the 2d Division; that they retain their identity, for morale's sake, as the 4th Brigade, United States Marine Corps. Pershing saw no harm in this as he had already decided that it would be the 4th Brigade of the 2d Division anyway. He agreed with Barnett and the Marines were officially designated as the 4th Brigade.

This identification caused a little more jealousy with some members of the Army than many realized. The situation was not helped when the Marines were compelled to trade in their forest-green uniforms for new Army khakis. As the new uniforms were distributed, the Marines cut off the Army buttons and sewed on their own Marine buttons. They did the same with their Marine insignia—they left no doubt or confusion as to who they were.

When the brigade was finally sent out for training with its Army comrades-in-arms and French instructors, another niggling point was introduced. The Marines had to exchange their trusty and reliable Lewis machine guns for Hotchkiss machine guns and French Chauchat automatic rifles. Both of these weapons were noted for their unreliability, and they were unusually heavy. Add to that the fact that the Marines had to train and learn Army drill, which was quite different from their own, in the middle of a bad European winter. They were billeted out in ramshackle, windy old sheds and barns with French peasants. They had colds, body lice, and a host of other maladies—and they had not even *seen* the front lines. When well-trained and disciplined troops with an esprit de corps are subjected to conditions like that, they turn mean, but they stay controlled.

With the ever-increasing threat of a major German assault, the Allies were starting to press Pershing to bring the

American Expeditionary Forces to the front lines. Pershing was determined that he was not going to have his Army split up and scattered in whatever holes and corners the Allies wished. Neither did he want his forces stuck in the trenches. He wanted them to be used in field warfare and he wanted them to fight alongside one another.

The Allies vacillated, but in March of 1918, Pershing did send troops to the trenches for training under fire. The 4th Brigade was sent to one of the quiet areas to the southeast of Verdun for two months. It was here that the first member of the 4th Brigade was killed by a shell burst on April 1, 1918. A few weeks later, forty Marines died in one mustard-gas attack.

The Marine Corps was at last in the fight, honing some of their battle skills and learning something of the nature of the enemy.

The brigade was pulled out of the trenches in the middle of May and sent on field warfare exercises. It had an unwelcome change of command as Brigadier General Doyen became very ill and Pershing, rightfully, sent him back home. He also instructed commandant Barnett that he need not send another Marine general out to replace him. Instead, Pershing appointed his own chief of staff, Brigadier General James Harbord, to take command of the 4th Brigade.

Pershing told Harbord that in his opinion the Marine brigade was the "most military" of the expeditionary forces, and that he, Harbord, would have no excuse for failure while commanding it. Harbord commented that if he did well with the brigade, he would have no idea how he did it. In his first meeting with the 5th and 6th's regimental commanders, Colonel Neville and Colonel Catlin, it is reputed that Harbord was reminded that the Marine Corps's motto was *Semper Fideles*, and that guaranteed the brigade's loyalty—even to an Army officer.

However, the change in command was seen by the Marines as still another indication that Pershing and his staff thought little of the Marine Corps. It made them more determined than ever to prove themselves in battle.

By this time, the German army had broken through in

the north near the Somme area and had driven two British armies back. The French howled at Pershing to release all his forces to accompany their 1st Army reserves to the north. Pershing was extremely reluctant to place any of his troops under the control of leaders who had demonstrated their total lack of ability in warfare over the past three years. But he compromised and sent the 1st Division to the north.

The Germans then launched a massive assault against the weak and ragged French Sixth Army in the northeast. The 6th Army had been very badly deployed on a ridge of hills to the north of the Aisne River, and the Germans had opened the offensive with an incredibly accurate and devastating artillery bombardment. When the bombardment ceased, eighteen German divisions poured across the hills and over the Aisne.

On the first day of the attack, the Germans advanced fourteen miles and the German high command was elated with its success. The bedraggled Sixth Army, a mixture of British and French forces, managed to slow down the advanced by just a fraction over the next three days. But the Germans had covered nearly forty miles and they were just a few miles to the northwest of the town of Château-Thierry. The German general Erich Ludendorff had stopped the advance—his troops were almost exhausted and supply problems would become acute if he did not consolidate the territory he had just taken. A few miles to the south of the German holding position was the main road from Metz to Paris. Ludendorff's troops were only fifty miles from the French capital and the Allies were certain that Paris was the goal of this assault.

It was now Sunday, May 30, and the 2d Division was resting after its field exercises. Orders were received to move out to the south and join the battered French Sixth Army which was fighting a rearguard action to the northwest of Château-Thierry.

After a grueling and totally infuriating forty-eight-hour period in the hands of the French transport elements, the tired and hungry 4th Brigade reached its assigned position near Château-Thierry. The Marines were right behind the French lines to the south of the small town of Torcy. Between the Marines and the town lay a small dark woods called *Bois de Belleau*—Belleau Wood.

The main German front line had not changed for three days, but raiding parties had been sent out and they had continued to harass the French troops. The 2d Division was now in position on either side of the Marines and was strung out in defensive formations on the Metz–Paris road. The artillery, which had at first been delayed by some mix-up with the French authorities, was finally hauled into the area and set up.

The Germans attacked from their consolidated positions in corps strength and easily pushed the French Army back toward the 2d Division. Directly in front of the Marine position, two German divisions stormed through the town of Torcy and occupied Belleau Wood.

The French were in full retreat, apparently under orders, and they streamed rearward through the Marine position. One French officer informed Marine Captain Lloyd Williams that a withdrawal had been called and that he, too, should retreat. Captain Williams's reply to the French officer is now in the history books: "Retreat, hell! We just got here!" The amazed French officer shook his head and started toward the rear.

The German army remained in Belleau Wood overnight and the following morning they started toward the Marines.

As the gray-clad infantry battalions emerged from the trees and started across the wheat fields, the Marines called in artillery fire. The Marine infantry opened fire from a range of about eight hundred yards. French and British observers watched in awe as the Marine riflemen demonstrated their marksmanship. Every Marine picked a target and felled it, then worked the bolt of his 1903 Springfield and drove another round into the breech, took quick and careful aim, and felled another enemy—again and again.

The Germans were hardly out of the woods and they were being shot down in rows. They were stunned and confused, yet they still marched forward; they could not see the Marines. One amazed British officer remarked that it was like watching hundreds of snipers in action. To the Marines, it was just like being on the firing ranges at Parris Island.

The finest fighting armies in Europe were being given a demonstration of battlefield marksmanship at its best. The Marines were the only fighting force in the world who were taught to select an individual enemy target and then put it down. Incredible as it may seem, all other fighting forces were taught to fire "in the path of the advancing enemy as rapidly as possible and at about knee height." The Marines' emphasis on marksmanship training and the time each individual Marine spent on the practice ranges far exceeded that of any other army in the world—including the U.S. Army. The Marines' selection of the Springfield rifle, and every rifle since that time, has been the result of a long and careful evaluation procedure. To this day the individual Marine still spends more time on the firing ranges than any of his contemporaries in other branches of the armed forces. The "shooting gallery of Belleau Wood," as one Marine sergeant put it, was to change the thinking of the infantry doctrine of all the armies of Europe from then on.

The German advance from Belleau Wood on that morning in June of 1918 was halted by the Marine riflemen and machine gunners with the aid of well-directed artillery. As a result, Ludendorff's army went to tactical defense positions along its front line.

The French commander, General Degoutte, was beside himself with delight at the success of the 4th Brigade. He had at last found a weapon which could take him out of the nightmare of the past three years—the Marine brigade. He immediately decided to mount an attack using the 2d Division, reinforced with the French 167th Division. He was concerned that the Germans would start to dig in and another bloody trench war would ensue, or that they would further reinforce and commence another assault.

The Marines were to assume the responsibility for the main attack with the 23rd U.S. Infantry on its right and a French regiment on its left. But the French general's timetable was not quite what it should have been, and the 4th Brigade and the assault battalions only received their orders a few hours before the attack.

The 5th Regiment opened the attack on Hill 142,

which commanded the ground around the town of Torcy. Artillery support was totally inadequate and, because of the short time given to prepare, only two companies started the first assault on the hill. As the Marines waded through fields of wheat and poppies toward the hill, the German machine gunners started to cut them down. The remainder of the brigade started toward the woods in the early afternoon. Once again artillery support was lacking and the skillful German machine gunners, firing from concealed positions within the woods, took a murderous toll.

In the woods, things were no better. Across one section alone, almost 1,000 Marines started toward the trees and only 550 got there. Once inside the dark woods, they fought machine-gun nest after machine-gun nest. Then they would get driven back and the nests would be reoccupied. Small clearings inside the woods were strewn with rocks and boulders, almost ideal conditions from which to conduct a defense, and the German troops were masters of the art.

The Marines around Hill 142 almost took Torcy, but were driven out before dark and the battle continued on the hill with both sides near exhaustion. The ferocity with which the Germans fought for the hill seemed to indicate that they were intent on keeping it, but the Marines had no intention of letting them have it back.

There was another small town called Bouresches on the south side of the woods. It had been taken by the Germans when they had first driven the French back. The Marines drove them out of the town by dusk, but they came back that evening and some terrible street fighting took place throughout the night.

Casualties on both sides were high on that first day, before darkness fell on Hill 142 and Belleau Wood. The United States Marine Corps had lost more men in a single day than it had in its entire history! The Marines held the hill, but the fighting in the woods continued, with bloody and vicious skirmishes mixed in with terrifying artillery barrages. After twelve days of attack and counterattack, the 4th Brigade had been so badly depleted and torn that it had to be withdrawn from the line. The U.S. 7th Infantry took over the battle while the 4th retired to rebuild.

On June 25 they were back in the woods again, among the blackened stumps, rocks, boulders, and machine-gun nests. In a coordinated, two-battalion attack, the 4th Brigade finally cleared the enemy out and the signal was sent to General Pershing: "Belleau Wood now U.S. Marine Corps entirely."

The battle for Belleau Wood had cost the 4th Brigade 112 officers and 4,598 men; of the casualties, there were more than 1,000 dead.

The French, British, and other military observers of the battle were truly impressed by the Marines' performance and by the 2d Division as a whole. All the 2d Division's infantry brigades were awarded unit citations for gallantry by the French. But perhaps the greatest compliment to the 4th Brigade's fighting ability was the unofficial one paid to it by its adversary, the German army. In the official reports concerning the battle, they constantly referred to the ferocity of the Marines, whom they called *"Teufelhunden,"* or "devil-dogs."

The French people displayed their appreciation when they renamed *Bois de Belleau*. Today it is called *Bois de la Brigade de Marine*, the "Wood of the Marine Brigade."

5
MARINE CORPS TRAINING

Boot camp. That is where it all starts. It is not just a name that the Marine Corps has an exclusive right to; every branch of the armed forces has a boot camp—it is where you start from the bottom up. With the Marines it is tough but, as thousands of people have proven, it is not impossible.

To become a Marine you must have a certain level of physical fitness, along with a certain type of character. Some people just do not have the physical abilities of others, and no amount of training can change that. If you were six feet tall with excellent facial features and had seventeen university degrees, but you weighed only sixty-one pounds, there is very little chance that you would make it as a Marine, unless it was possible to instantly put another ninety or so pounds on your body.

But most people would meet the Marine Corps's basic physical requirements. If you have the ability and character from which the Marine Corps can build a good Marine, then you will be accepted into boot camp. But just because you get into boot camp, there is no certainty that you will come out as a Marine. It is possible that you will just not be able to do what you thought you could do; it's also possible that your Marine Corps instructors may discover that you could not do what they thought you could do!

If you cannot progress and meet the various requirements as they come along, you will not make it into the

Marines. Marine Corps instructors are there to help you train, not just to make you train. If they see that you are having problems in a certain area they will offer assistance, if assistance is all that is required.

During the initial screening of applicants, the recruiters and instructors attempt to ensure that prospective Marines have the necessary abilities. They are not going to waste your time, or their own, if it is obvious from the start that you cannot make it or do not want to make it.

When a recruit has to leave boot camp because he is not the right material, questions are asked by high authorities and they have to be answered. Questions such as: What happened? How did the individual get past the initial screening? Did someone make a mistake? The training staff simply does not like that sort of questioning, nobody does.

Marine Corps boot camp takes eleven weeks and there are four basic phases of training.

Phase one: The first phase is administrative processing, which takes three days. It starts with a haircut and a routine background investigation and is followed by issue of basic uniforms and toilet articles, medical examinations, and inoculations. This is followed by introductory training films and an introduction to the training staff personnel.

Phase two: With administrative processing completed, actual training starts. The training day begins at 5:00 A.M., Monday through Saturday. There is one hour free each night in which the recruit may write letters, read, or watch instructional television. Taps is at 9:00 P.M., and reveille on Sundays and holidays is at 6:00 A.M. On Sundays, religious services are held in the morning and organized athletics in the afternoon.

During the first few weeks of boot camp, the recruit learns how to make his rack, keep his squad bay clean, wear his uniform correctly, and drill. The drill instructors see to it that recruits get a lot of exercise. Classes will be attended on the history of the Marines, military courtesy and discipline, the M16A1 rifle, close order drill, first aid, close combat, and various other subjects. Tests will be taken, the recruit will be issued a rifle, and many obstacle courses will be run.

Phase three: The next training phase starts with visits to the rifle range, where the recruit will be taught how to shoot at and hit targets. During the first week on the rifle range safety procedures and firing techniques will be taught. Then a full week of firing the rifle will commence and end with qualification tests. There are three levels: marksman, sharpshooter, and expert. The Marines prefer experts. The final week of the second phase brings a change from the normal training routine as recruits serve as members of the work force in one of the dining facilities and provide maintenance for other depot facilities.

Phase four: The final phase starts after the mess and maintenance week. There is more drill and there is guard duty. Service uniforms will be fitted and tailored, and there will be more close combat training and more classroom work. There will be individual combat training and an amphibious and helicopter assault indoctrination. During the final days of boot camp, there is a drill evaluation and the final physical fitness testing. The recruit then gets paid and prepares for graduation. In those final days the recruit will be assigned his choice of a Military Occupational Specialty (MOS).

After graduation, the recruit will get ten to fifteen days leave before reporting to his next duty station. It is at this first duty station after boot camp that the Marine commences his training in his MOS: infantry, field artillery, operational communications, motor transport, ordnance, tank and amphibian tractor, and so forth. The list of opportunities is long, and the length of time a Marine spends in occupational training will be determined by the specialty (see list on page 42). Once this training has been completed, the Marine will be assigned to an operational unit.

Training never stops. In every discipline there are new developments and advances and, to be effective in its function, the Marine Corps must stay current and practiced. Encouragement and assistance is given to anyone who wishes to further his or her education, and there are active scholarship programs available.

Training and practicing in the individual skills of any one element is important, but there comes a time when battlefield skills must be practiced in the field with the various elements working together. The Marines place a great deal of emphasis on practical training—experience has shown that it pays off.

Fleet exercises with amphibious assaults are a regular part of the Fleet Marine Force's training. A variety of beaches are chosen and everyone from the Navy ship's captain to the Marine infantryman gets to practice the skills of his or her profession under conditions simulating an actual battle. When it comes to realistic military training, the Marine Corps is ahead of most. This is perhaps best demonstrated in the hill and desert areas around the Marine Corps's Air-Ground Combat Center at Twenty-nine Palms, California.

Battalion-, brigade-, and regimental-size exercises are carried out in the vast and rugged wilderness of the California high desert. Some of the exercises last for over a month; during that time the Marines go without seeing anything that resembles civilization, except for the lights of distant cities at night. During the summer days the sun brings the temperature up to 125 degrees Fahrenheit, and at night the temperature drops to the high 30s. In winter the high daytime temperatures are in the mid 80s, and the nighttime lows sometimes come down to zero. There is almost always wind and the inevitable sandstorms; in winter there is rain and the ensuing mud.

The Air-Ground Combat Center is the only military installation in the Western world that allows for maneuvering through the ordnance impact areas. Exercise is on a regular basis with the various marine units shipping into the area from their base and fleet locations.

During the exercise, a battalion faces a Soviet-type threat consisting of a reinforced motorized rifle battalion. A Marine amphibious brigade would be pitted against a motorized rifle regiment.

As the exercise force attacks, the enemy initially defends and then commences a covered withdrawal. Initial information on the enemy is distributed in a letter of instruction and then in the form of intelligence summaries. Once the exercise has commenced, the enemy is simulated by means of input from the Tactical Exercise Evaluation and Control Group, the Direct Air Support Center, and various unit controllers. The enemy is also represented on the ground by specially constructed targets which must be eliminated by fire.

Although the emphasis of the exercise is on the ability of the commander and his staff to coordinate supporting arms, all other aspects—field artillery, infantry, armor, missiles—are thoroughly evaluated.

The Marines are readied behind a line of departure and, as H-hour approaches, simulated naval gunfire commences from behind the waiting troops. Mobile 8-inch guns are set to the extreme rear and towed 155-mm guns are positioned ahead of them; both are firing directly over the heads of the waiting troops.

High performance jet aircraft sweep in between the artillery barrages and saturate the area with high explosives and mixed ordnance. Mortar crews along the line of departure keep up a steady bombardment together with the artillery and aircraft. A tremendous smoke screen is then laid by a high speed aircraft and the mobile assault companies attack across the line of departure.

Everyone moves forward in order. Tanks engage individual targets with their main guns, helicopters work on the flanks, and the artillery and jet aircraft continue to pound the area some 2,000 meters ahead of the assault units. Any observer who has experienced combat will readily confirm that the sights and sounds of battle are at a level very similar to actual combat.

The artillery starts to walk its barrage back, attack helicopters and missile-firing vehicles join the fray. As the enemy is driven back, the command post moves forward and the commander now starts to face the challenge of maintaining command and control. The artillery units start to move their 155s forward into new positions, reserve forces and the logistic elements are also moved forward. The logistic challenges are real because of the terrain, the climate, and the live-fire nature of the exercise. Ammunition, food, water, and equipment are moved forward. Simulated (and actual) casualties are evacuated to the rear. Ordnance engineers are everywhere, clearing simulated minefields and handling unexploded rounds. The infantry assaults selected positions with support by tanks and mortars. The air is filled with machine-gun and tracer fire, white phosphorus, the smell of napalm, diesel, and CS (tear) gas. The whistle of artillery shells and the general melee of battle are everywhere. There is attack and counterattack.

It continues like this for three days and three nights. By the end of the exercise, the Marines will have moved forward some twenty miles from the starting point. It is as close to combat conditions as possible. It is realistic training, carefully designed to prepare the Marine Corps for the battlefield, should they have to fight again.

Military Occupational Specialties of the Marines

The following lists specialties, the time it takes to train, and the location of the training school.

*Asterisk indicates an advanced school.
Women Marines may be assigned into all occupational fields except Field Artillery (OCC Fld 08), Infantry (OCC Fld 03) and Tank and Amphibian Tractor (OCC Fld 18).

Air Control/Air Support/Anti-Air Warfare: Occupational Field 72

Air Command/Control Electronics	6 weeks	29 Palms, CA
Air Support Operations Operator	14 weeks	29 Palms, CA
Hawk Missile System Operator	8 weeks	Ft. Bliss, TX
Tactical Air Controller	5 weeks	29 Palms, CA
Redeye Gunner	6 weeks	Ft. Bliss, TX

Aircraft Maintenance: Occupational Field 60/61

Aviation Maintenance Administration	7 weeks	Meridian, MS
Aviation Machinist Mate (Jet Engine)	7 weeks	Memphis, TN
Air Crew Survival Equipment Man	18 weeks	Lakehurst, NJ

Aviation Safety Equipment	8 weeks	Memphis, TN
Aviation Structure Mechanic	9 weeks	Memphis, TN
Aviation Hydraulic Mechanic	7 weeks	Memphis, TN
Cryogenics Operator	14 weeks	Portsmouth, VA
Aviation Support Equipment Technician (Mechanical)	9 weeks	Memphis, TN
Aviation Support Equipment Technician (Hydraulic)	9 weeks	Memphis, TN
Aviation Support Equipment Technician (Electrical)	9 weeks	Memphis, TN
Basic Helicopter Course	6 weeks	Memphis, TN

Airfield Services: Occupational Field 70

Aviation Operations Clerk	6 weeks	Meridian, MS
Expeditionary Airfield Technician	9 weeks	Lakehurst, NJ
Aircraft Firefighting/Rescue	6 weeks	Memphis, TN

Air Traffic Control & Enlisted Flight Crews: Occupational Field 73

Aerial Navigator	24 weeks	Mather AFB, CA
Airborne Radio Operator	13 weeks	Cherry Point, NC
Air Traffic Controller	19 weeks	Memphis, TN
*Air Support Operations Operator	14 weeks	29 Palms, CA

Ammunition & Explosive Ordnance Disposal: Occupational Field 23

Ammunition Storage	6 weeks	Redstone Arsenal, AL
Basic Surface/Nuclear Weapons Disposal	14 weeks	Indianhead, MD
Surface Explosive Ordnance Refresher Course	2 weeks, 4 days	Indianhead, MD

Auditing, Finance & Accounting: Occupational Field 34

Personal Financial Records Clerk	9 weeks	Camp Lejeune, NC

*Advanced Disbursing	11 weeks	Camp Lejeune, NC
Travel Clerk	7 weeks	Camp Lejeune, NC
Fiscal Accounting	9 weeks	Camp Lejeune, NC

Aviation Ordnance: Occupational Field 65

Aviation Ordnance Munitions Technician	9 weeks	Memphis, TN
*Aviation Ordnanceman Advanced	21 weeks	Memphis, TN

Avionics: Occupational Field 63/64

Avionics Technician	18 weeks	Memphis, TN
Advanced First Term Avionics	20 weeks	Memphis, TN
Aviation Electrician Mate	11 weeks	Memphis, TN
*Aviation Electrician Intermediate	25 weeks	Memphis, TN
Precision Measuring Equipment Technician	29 weeks	Lowry AFB, CO
*Avionics Technician Intermediate	29 weeks	Memphis, TN

Band: Occupational Field 55

Drum & Bugle Corps	40 weeks	Little Creek, VA
Music, Basic, Class A	24 weeks	Little Creek, VA
*Music, Intermediate, Class C	23 weeks	Little Creek, VA
*Music, Assistant Bandleader, Class C1	40 weeks	Little Creek, VA

Data/Communications Maintenance: Occupational Field 28

Fire Control Computer Repair	10 weeks	29 Palms, CA
	10 weeks	Aberdeen, MD
	29 weeks	Ft. Bliss, TX
Artillery Electronic Equipment Repair	12 weeks	29 Palms, CA
Weapons Location Equipment Repair	5 weeks	29 Palms, CA
Basic Electronics	14 weeks	29 Palms, CA
Teletype Repair	15 weeks	29 Palms, CA
Field General Comsec Repair	32 weeks	Ft. Gordon, GA
Fixed Ciphony Repair	35 weeks	Ft. Gordon, GA
*Aviation Radio Technician	11 weeks	29 Palms, CA

U.S. MARINES

*TSEC/KY-8 Maintenance (Cryptologic Repair)	8 weeks	Vallejo, CA
*Terminal Equipment Theory	5 weeks	29 Palms, CA
Microwave Equipment Repair	3 weeks	29 Palms, CA & Keesler AFB, MS
Ground Radio Repairer	18 weeks	29 Palms, CA
Test Instrument Repair	21 weeks	Albany, NY
Telephone Switchboard Repair	10 weeks	29 Palms, CA
Cable Splicing Specialist	5 weeks	Sheppard AFB, TX
Aviation Radio Repairer	11 weeks	29 Palms, CA
Meteorological Equipment Maintenance	16 weeks	Chanute AFB, IL

Data Systems: Occupational Field 40

Electrical Accounting Machine Operator	5 weeks	San Diego, CA
IBM System/360 Operations System (OS) Operation Course	4 weeks	Quantico, VA
IBM System/360 Operating System (OS) COBOL Programming Course	8 weeks	Quantico, VA
IBM System/360 Operating System Programmer Course	9 weeks	Quantico, VA
IBM System/360 Operating System (OS) Advanced Programming Techniques Course	6 weeks	Quantico, VA
IBM System/360 Operating System (OS) Data Control Techniques	6 weeks	Quantico, VA
*Data Processing Installation Management Seminar	4 weeks	Quantico, VA
*Advanced Mark IV File Management System Course	2 weeks	Quantico, VA

Drafting, Surveying, Mapping: Occupational Field 14

Basic Cartography	11 weeks	Ft. Belvoir, VA
Construction Surveying	15 weeks	Ft. Belvoir, VA
Geodetic Surveying	15 weeks	Ft. Belvoir, VA
Construction Drafting	11 weeks	Ft. Belvoir, VA
Photogrammetric Compilation	14 weeks	Ft. Belvoir, VA

Electronics Maintenance:
Occupational Field 59

Aviation Radio Technician	11 weeks	29 Palms, CA
Aviation Radio Repairer	12 weeks	29 Palms, CA
Meteorological Equipment Maintenance	16 weeks	Chanute AFB, IL
Improved Hawk Fire Control Technician	29 weeks	Ft. Bliss, TX
Improved Hawk Information Coordination Central Maintenance	25 weeks	Ft. Bliss, TX
Improved Hawk Firing Section Repairer	24 weeks	Ft. Bliss, TX
*Improved Hawk Missile System Maintenance Technician	39 weeks	Redstone Arsenal, AL
Improved Hawk Mechanical System Repairer	32 weeks	Redstone Arsenal, AL
*Improved Hawk Simulator System Technician	8 weeks	Ft. Bliss, TX
*Improved Hawk Pulse Radar Technician	35 weeks	Redstone Arsenal, AL
*Improved Hawk Continuous Wave Radar Technician	23 weeks	Redstone Arsenal, AL
*Improved Hawk Fire Control Technician	42 weeks	Redstone Arsenal, AL
*Aviation Radar Technician	19 weeks	29 Palms, CA
*Ground Radar Technician	22 weeks	29 Palms, CA
*Aviation Fire Control Repair	10 weeks	29 Palms, CA
Tactical Air Operations Central Repairer	18 weeks	29 Palms, CA
*Tactical Air Operations Central Technician	35 weeks	29 Palms, CA
Tactical Air Command Central Technician	22 weeks	29 Palms, CA
*Tactical Data Communications Central Technician	20 weeks	29 Palms, CA
*Tactical Air Command Central Repairer	17 weeks	29 Palms, CA
*Tactical Data Communications Central Repairer	16 weeks	29 Palms, CA
Ground Radar Repair	7 weeks	29 Palms, CA
Aviation Fire Control Technician	8 weeks	29 Palms, CA
Aviation Radar Repairer "B"	11 weeks	29 Palms, CA
Aviation Radar Repairer "C"	12 weeks	29 Palms, CA

*Marine Air Traffic Control Unit Navigational Aids Technician	38 weeks	Memphis, TN
*Marine Air Traffic Control Unit Communications Technician	29 weeks	Memphis, TN
*Marine Air Traffic Control Unit Radar Technician	33 weeks	Memphis, TN
Marine Air Traffic Control Unit Navigational Aids Repair	32 weeks	Memphis, TN
Marine Air Traffic Control Unit Radar Repairer	11 weeks	Memphis, TN
Marine Air Traffic Control Unit Communications Repair	11 weeks	Memphis, TN
Tactical Data Systems Maintenance Technician	34 weeks	Mare Island, CA
*Tactical General Purpose Computer	24 weeks	29 Palms, CA

Engineer, Construction, Equipment & Shore Party: Occupational Field 13

Basic Combat Engineer	7 weeks	Camp Lejeune, NC
Engineer Equipment Mechanic	11 weeks	Camp Lejeune, NC
Basic Metal Worker	12 weeks	Camp Lejeune, NC
Engineer Equipment Operator	9 weeks	Ft. Leonard Wood, MD
*Engineer Operations Chief	12 weeks	Camp Lejeune, NC
*Journeyman Combat Engineer	11 weeks	Camp Lejeune, NC
*Petroleum Products Analysis	12 weeks	Ft. Lee, VA
*Journeyman Metal Worker	6 weeks	Camp Lejeune, NC
*Journeyman Shore Party Specialist	4 weeks	Camp Lejeune, NC
*Soils Analysis	8 weeks	Ft. Belvoir, VA

Field Artillery: Occupational Field 08

Shore Fire Control Party	4 weeks	San Diego, CA & Norfolk, VA

THE FIGHTING ELITE

Field Artillery Fire Controlman	7 weeks	Ft. Sill, OK
Field Artillery Scout Observer	4 weeks, 2 days	Ft. Sill, OK
Field Artillery Meteorological Crewman	8 weeks	Ft. Sill, OK
Field Artillery Radar Crewman	6 weeks	Ft. Sill, OK
Meteorological Organizational Maintenance Repairer	10 weeks	Ft. Sill, OK

Food Service: Occupational Field 33

Basic Bakers Course	8 weeks	Camp Lejeune, NC
Basic Food Service Course	6 weeks	Camp Lejeune, NC
*Food Service Staff NCO Leadership	9 weeks	Camp Lejeune, NC
*Food Service NCO Leadership	12 weeks	Camp Lejeune, NC
*Bakery NCO Leadership	11 weeks	Camp Lejeune, NC

Infantry: Occupational Field 03

Rifleman	4 weeks	Camp Pendleton, CA Camp Lejeune, NC
Machine Gunner	4 weeks	Camp Pendleton, CA Camp Lejeune, NC
Mortar Man	4 weeks	Camp Pendleton, CA Camp Lejeune, NC
Antitank Assaultman	4 weeks	Camp Pendleton, CA Camp Lejeune, NC

U.S. MARINES

Intelligence: Occupational Field 02

Image Interpretation—Officer Basic Intelligence	12 weeks	Ft. Huachuca, AZ
Amphibious Intelligence (Entry Level)	4 weeks	Norfolk, VA & Coronado, CA
*Amphibious Intelligence (Intermediate Level)	3 weeks	Norfolk, VA & Coronado, CA
*Amphibious Intelligence Specialist (Advanced)	4 weeks	Norfolk, VA

Legal Services: Occupational Field 44

*Advanced Legal Services	10 days	Camp Pendleton, CA
Legal Services Specialist	5 weeks	Camp Pendleton, CA
Legal Services Reporter Notereader/Transcriber	13 weeks	Camp Pendleton, CA

Logistics: Occupational Field 04

Landing Ship Embarkation	2 weeks	San Diego, CA & Norfolk, VA

Marine Corps Exchange: Occupational Field 41

*Armed Forces Culinary Course	2 weeks	Patuxent River, MD
Principles of Cost Controls	5 days	Patuxent River, MD
*Management of Marine Corps Clubs	4 weeks	Patuxent River, MD

Military Police & Corrections: Occupational Field 58

*Criminal Investigation	15 weeks	Ft. McClellan, AL

Law Enforcement (Military Police)	8 weeks	Lackland AFB, TX
Law Enforcement (Correctional Specialist)	7 weeks	Lackland AFB, TX
*Law Enforcement (NCO Advanced Course)	10 weeks	Ft. McClellan, AL

Motor Transport: Occupational Field 35

Fuel & Electrical Systems Repair	11 weeks, 4 days	Aberdeen, MD
Basic Automotive Mechanic	13 weeks	Camp Lejeune, NC
Vehicle Body Repair	10 weeks	Aberdeen, MD
*Advanced Automotive Mechanic	17 weeks	Camp Lejeune, NC
Motor Transport Staff NCO Course	8 weeks	Camp Lejeune, NC

Nuclear, Biological & Chemical: Occupational Field 57

CBR Enlisted	5 weeks	Ft. McClellan, AL
*Nuclear Emergency Team	6 weeks	Kirtland AFB, NM

Operational Communications: Occupational Field 25

Field Radio Operator	9 weeks	29 Palms, CA
Radioman, Class A	16 weeks	Norfolk, VA
Communications Center Operator	11 weeks	29 Palms, CA
Microwave Equipment Operator	4 weeks	29 Palms, CA
*Operational Communications Chief	19 weeks	29 Palms, CA
*Radio Chief	13 weeks	29 Palms, CA
*Wire Chief	14 weeks	29 Palms, CA
High Frequency Communications Central Operator	5 weeks	29 Palms, CA

U.S. MARINES

Ordnance: Occupational Field 21

Assault Amphibian Repairer	8 weeks	Camp Pendleton, CA
Tracked Vehicle Repairer, Self-Propelled Artillery	5 weeks	Aberdeen, MD
Tracked Vehicle Repairer Tank Course	6 weeks	Aberdeen, MD
Small Arms Repair	6 weeks	Aberdeen, MD
Optical Instrument Repair	14 weeks	Aberdeen, MD
Machinist	15 weeks	Aberdeen, MD
*Turret Repairer	8 weeks	Aberdeen, MD
Artillery Repair	12 weeks	Aberdeen, MD

Personnel & Administration: Occupational Field 01

Administrative Clerk	8 weeks	Camp Lejeune, NC & Camp Pendleton, CA
Unit Diary Clerk	8 weeks	Camp Lejeune, NC & Camp Pendleton, CA
Personnel Clerk	8 weeks	Camp Lejeune, NC & Camp Pendleton, CA

Printing & Reproduction: Occupational Field 15

USMC Basic Offset Printing	9 weeks	Ft. Belvoir, VA
Basic Photolithographic Process	14 weeks	Ft. Belvoir, VA

Public Affairs: Occupational Field 43

Basic Broadcaster	10 weeks	Indianapolis, IN
Basic Journalist	10 weeks	Indianapolis, IN

52 THE FIGHTING ELITE

*Newspaper Editor	3 weeks	Indianapolis, IN
*Public Affairs Supervisor	3 weeks	Indianapolis, IN

Signals Intelligence/Ground Electronic Warfare: Occupational Field 26

Radio Direction Finder/Electronic Warfare Operator	5 weeks	Pensacola, FL
Non-Morse Intercept/Electronic Warfare Operator	20 weeks	Pensacola, FL
Foreign Language Training	24-47 weeks	Monterey, CA
Cryptologic Technician "O," Class A	14 weeks	Pensacola, FL
Cryptologic Technician "R," Class A	22 weeks	Pensacola, FL
Cryptologic Technician "T," Class A Preparatory	12 weeks	Pensacola, FL
Manual Morse Intercept Electronic Warfare Operator	20 weeks	Pensacola, FL
*Voice Processing Specialist	12-15 weeks	San Angelo, TX
*Special Intelligence Communication Center	18 weeks	Pensacola, FL

Supply Administration & Operations: Occupational Field 30

Basic Supply Stock Control	7 weeks	Camp Lejeune, NC
Marine Aviation Supply	13 weeks	Meridian, MS
Senior Enlisted Aviation Supply Management	4 weeks	Meridian, MS
Subsistence Supply Clerk	5 weeks	Camp Lejeune, NC
*Supply Chief Leadership	6 weeks	Camp Lejeune, NC
*Supply NCO Leadership	11 weeks	Camp Lejeune, NC
*Warehousing SNCO Leadership	5 weeks	Camp Lejeune, NC
Defense Basic Preservation/Defense Vehicle Processing for Shipment/Storage	3 weeks	Albany, GA

Tank & Amphibian Tractor: Occupational Field 18

*Assault Amphibian Unit Leader	5 weeks	Camp Pendleton, CA
Armor Crewman, Advanced Individual Training	9 weeks	Ft. Knox, KY
Amphibian Crewman Course	5 weeks	Camp Pendleton, CA

Training and Audio Visual Support: Occupational Field 46

Audio/Television Production Specialist	7 weeks	Lowry Technical Training Center, CO
Graphics Specialist	12 weeks	Lowry Technical Training Center, CO
Still Photographer Specialist	5 weeks	Lowry Technical Training Center, CO

Transportation: Occupational Field 31

Joint Personal Property	2 weeks	Ft. Eustis, VA
*Defense Advanced Traffic Management	3 weeks	Ft. Eustis, VA
Installation Traffic Management	4 weeks	Ft. Eustis, VA

Utilities: Occupational Field 11

Basic Refrigeration Mechanic	7 weeks	Camp Lejeune, NC
Basic Hygiene Equipment Operator	13 weeks	Camp Lejeune, NC
Basic Electrician	6 weeks	Camp Lejeune, NC
*Journeyman Electrician	12 weeks	Camp Lejeune, NC
*Journeyman Hygiene Equipment Operator	15 weeks	Camp Lejeune, NC

THE FIGHTING ELITE

Electrical Equipment Repair Specialist	16 weeks	Camp Lejeune, NC
Utilities Chief	18 weeks	Camp Lejeune, NC
Office Machine Repair Specialist	15 weeks	Ft. Lee, VA
Fabric Repair Specialist	7 weeks	Ft. Lee, VA

Weather Service: Occupational Field 68

Weather Observer Aerographer Mate—Class A	11 weeks	Chanute AFB, IL
*Weather Forecaster	19 weeks	Chanute AFB, IL
Rawinsonde/Radiosonde Operator	6 weeks	Chanute AFB, IL

Special Field Training

*Career Information & Counseling	4 weeks	San Diego, CA
*Recruiter	7 weeks	San Diego, CA
*Drill Instructor	8 weeks	Parris Island, SC & San Diego, CA
Marine Security Guard	6 weeks	Quantico, VA
*Sea Duty Indoctrination	4 weeks	San Diego, CA
Memorial Activities Chapel Management Specialist	4 weeks	Keesler AFB, MS
Staff NCO Academy	6 weeks	Quantico, VA; El Toro, CA & Camp Lejeune, NC

Chart taken from *U.S.M.C. Guide for Counsellors and Advisors of Students*.

6

MARINE INFANTRY
Small Arms and Squad Deployment

Despite the incredible technological advances in the weapons of war since the time of Napoleon, the undisputed queen of battle is still the infantry.

Perhaps it is artillery, helicopters, sophisticated fighter and bomber aircraft, tanks, armored vehicles, missiles, and other major and minor breakthrough weaponry that make it possible for an army to advance. But it is the infantry that must go forward to occupy the strategic positions that are the key to victory. Territory that is occupied must be organized and defended against counterattacks, which will invariably come. Both the initial advance and the subsequent holding of territory gained must depend upon the fire power of the infantry itself.

Throughout today's Marine Corps, the Marine infantrymen are referred to in seemingly derogatory terms—"grunts," "ground pounders," "gravel crunchers," to name a few. Yet those who cast the names will tell you quite clearly that their own function is to support and assist those "grunts" at all times and at all cost.

The Marine infantry is equipped with a surprising array of weapons for increasing its fire power, including light and heavy machine guns, mortars, grenade launchers, anti-tank and anti-aircraft rockets with wire, and infra-red and laser guidance systems. These are very powerful auxiliaries, but they have never taken over the duties of the individual marksmen who make up the larger part of the infantry battalion. In the final analysis, it is the skill of these individuals that will most likely bring about the success, or failure, of any infantry operation.

From its inception the Marine Corps has recognized this fact, and it has served well. To this day, every Marine, regardless of status or sex, has been trained as a rifleman and has learned *The Creed of the United States Marines,* "My Rifle."

> This is my rifle. There are many like it, but this one is mine. My rifle is my best friend. It is my life. I must master it as I must master my life. My rifle, without me is useless. Without my rifle, I am useless. I must fire my rifle true. I must shoot straighter than my enemy who is trying to kill me. I must shoot him before he shoots me. I will . . . My rifle and myself know that what counts in war is not the rounds we fire, the noise of our burst, nor the smoke we make. We know that it is the hits that count. We will hit . . . My rifle is human, even as I, because it is my life. Thus, I will learn it as a brother. I will learn its weaknesses, its strength, its parts, its accessories, its sights, and its barrel. I will ever guard it against the ravages of weather and damage. I will keep my rifle clean and ready, even as I am clean and ready. We will become part of each other. We will . . . Before God I swear this creed. My rifle and myself are the defenders of my country. We are the masters of our enemy. We are the saviors of my life. So be it, until there is no enemy, but Peace!

Infantry Small Arms

The basic weapon of the Marine Corps is still the rifle. The current general issue rifle is the M16A1, more commonly referred to as the M16, which has evolved from a long line of proven rifles.

It was always thought that rifles had to be handmade to be of any substance. However, modern technology now permits mass production of rifles from lightweight alloys and plastics, and some of the current mass-produced rifles far surpass handmade guns of only a few years ago.

The M16 is one of the better examples of a massproduced rifle. It is a 5.56-mm, magazine-fed, gasoperated, air-cooled weapon. The aluminum magazine holds thirty rounds of ammunition and the total firing weight of the loaded weapon, with a sling installed, is just eight pounds. An M7 bayonet-knife can be mounted by attaching it to a stud directly below the front sight assembly.

Absolute maximum range is 2653 meters, but the maximum effective range is 460 meters. (The Marine Corps uses the metric system for all field ranges and distances. The principal reason for this is that most of the nations of the world use the metric system on their maps, road signs, and so forth. When working in conjunction with our allies, confusion over distances and ranges can have disastrous results.)

The M16 has a selector switch that gives a choice of one of two modes of fire: semi-automatic or automatic. Semi-automatic simply means that the gun, after the initial manual loading and cocking, will load itself after each round is fired and the trigger must be released and squeezed again to fire the next round. In the semi-automatic mode, the rate of fire is between 45 and 60 rounds a minute, depending upon how fast the individual Marine can squeeze the trigger and change magazines. In the fully automatic mode, the trigger is kept squeezed after cocking and the rate of fire is between 150 and 220 rounds a minute. Once again, this rate is dependent upon the operator, but this time it is governed only by the speed at which he can change magazines.

It must be remembered that it is not just speed that counts. Indiscriminate use of ammunition in a battlefield situation is foolhardy; there is a limited amount of ammunition that an individual can carry, and sometimes it can be difficult to get supplies of munitions to the front line. Shooting concentrated, purposeful, and well-directed fire is one of the commandments of the marine infantry.

The number-two weapon of the infantryman is the rifle and bayonet combination. In the hands of a skilled marine it becomes a multipurpose weapon to be used as spear, sword, shield, or club. At night this combination weapon can kill silently and with surprise. In hand-to-hand fighting, when the rifle cannot be reloaded and the use of grenades is impractical, it is the decisive weapon. At times like these, it is the practiced and aggressive bayonet fighter who will win, and it is for this reason that bayonet practice is an essential part of the infantryman's training.

The third weapon of importance to the infantryman is the M203, 40-mm grenade launcher. It is a lightweight, single-shot, breech-loaded, pump-action, shoulder-fired weapon. It is attached to the barrel of a slightly modified M16A1 rifle and does not interfere with the normal function of the rifle. The grenade launcher has its own trigger mechanism just ahead of the rifle's magazine, and a special set of sights is also attached to the top of the rifle. The total weight of the loaded rifle with the empty M203 attached is eleven pounds and it is an extremely useful weapon when used correctly. The maximum effective range of the launcher for general area coverage is 350 meters; for a point target it is 150 meters. In combat, it can be fired safely at targets as close as 31 meters.

The M203 is a very versatile weapon. It uses more types of rounds than any other single case weapon. Apart from many new types of rounds now under development and some specialty rounds which are already available, there are seven currently issued 40-mm rounds.

The most commonly used round is the high-explosive M406. It is classed as a "casualty producing round," with an effective hit radius of five meters. On impact, it explodes and fires out some 325 fragments. As a safety precaution, this round is "semi-smart." It does not arm until it travels between 14 and 28 meters away from the launcher, thus preventing a premature explosion close to friendly lines.

The second round, called HEDP—the High Explosive

Dual Purpose M433—is an anti-armor, anti-personnel round that can penetrate two inches of homogeneous armor. It, too, is semi-smart, and has a five-meter casualty radius.

The next type of round is a fearful little monster known as the Bounding Fragmentation round—the M397. This is not a "point detonating round," as it does not completely explode on impact. Instead, a small explosive charge fires the projectile approximately five feet back into the air where it then explodes, thus creating a much greater casualty radius than the M406 or the M433. This round is effective when used against troops in open areas or in fighting holes without overhead cover.

The fourth round is the M576, Multiprojectile Canister, which is filled with buckshot. It is most effective in close-quarter battle situations at a range of up to 30 meters and can be used in jungle environments, built-up areas, and in poor visibility, when targets are at close range.

The fifth round is the M651A1, Tactical CS. It has a maximum range of 400 feet and is point-detonating. After impact, it releases the nonpersistent CS gas for approximately twenty-five seconds.

The sixth and final combat round is the M583A1, White Star Parachute. It explodes about 180 meters in the air and, for about forty seconds, produces illumination of 45,000 candlepower.

The seventh and final round in general issue is the M407A1, Training round. It explodes on impact and disperses a yellow dye to mark the location of the hit.

Although not normally seen hanging on the hip of every Marine infantryman, the pistol or sidearm must be mentioned. Every Marine is taught how to use it as it is issued for certain duties where a rifle would be cumbersome or unwieldy. In combat situations it is more often carried by NCOs and commissioned officers, along with their rifles.

The current issue sidearm is the .45-caliber automatic pistol, the M1911A1. It is a recoil-operated, magazine-fed weapon with a magazine capacity of seven rounds. It has a maximum range of 1500 meters, but is only effective up to 50 meters. It is a close-quarter battle weapon only, and is almost at the end of its service life as the Marine Corps is actively trying to replace it with a more modern weapon.

The next weapon most familiar to the Marine infantryman is the machine gun. Machine guns are an American invention, and the patent office first described them as "being guns capable of sustained fire from energy derived from an outside source." The outside source in those days was a hand-operated crank. The first recorded use of such a gun was at the Battle of Fair Oaks, Virginia, on May 31, 1862. In this battle, the Confederate troops deployed a battery of Williams machine guns. The weapons had little real influence on the Civil War, having many drawbacks in design and manufacture. The first really effective machine guns were the designs of Richard Jordan Gatling and, to this day, his last name is still associated with the weapon.

Modern machine guns can deliver a very high rate of sustained and withering fire. With the Marines, they are normally used as a support weapon in the assault as they are light enough to be carried in the attack. But in defense, they are the backbone of the Marine unit. In a defensive situation, the guns are placed so as to cover all avenues of approach to the position being defended, and to deliver overlapping and interlocking bands of "grazing fire."

Each individual machine gun is normally set up and allocated a "sector of fire"; anything that is not friendly and moves within that sector is the responsibility of that gun. However, should the enemy launch an assault against the whole of the defensive position, then the machine guns shift their fire to "final protective lines." These lines are established so that every gun's fire overlaps one or more other gun's. In this fashion, several intersecting bands of grazing fire can be delivered across the front toward the approaching enemy, providing a lethal defensive screen. Barbed wire and mines can be set as obstacles to force enemy troops to channel into areas where the greatest concentration of fire can be directed.

Successful operation of a machine gun is vital, and marine training places a great deal of emphasis on efficient operation and marksmanship.

The present-day machine gun is the M60. It is a 7.6-mm, belt-fed, gas-operated, air-cooled, fully automatic weapon. The 7.62 rounds are fed into the gun by means of a disintegrating, metallic, split-link belt. The individual links of the belt are stripped off by the breech mechanism as each round is fed into the chamber and fired. Expanding gases from the fired rounds drive the bullets through the barrel and also provide the necessary energy to operate the gun.

Although the gun itself weighs only twenty-three pounds, a two-man crew operates the weapon—one to fire and one to load and feed the links. The second man is also needed to carry spare equipment and ammunition. Occasionally, a third man will be assigned to the crew in order that more ammunition may be carried.

The M60 comes complete with its own bipod mounted under the front of the barrel; a tripod mount, which weighs nineteen pounds, is also carried for ease of operation when the gun is set up in a more permanent location.

The maximum effective range of the M60 is 1100 meters, but the weapon is deadly up to 3725 meters. The weapon is very popular with Marine gunners because of its light weight, accuracy, and ease of maintenance.

The fifth and final piece of equipment which the marine infantryman is most familiar with is the hand grenade. It is the oldest type of weapon used by the American fighting man.

The origin of the hand grenade dates back to the Chinese before the eleventh century. Present historical evidence shows that the Chinese probably invented black powder in about A.D. 1000. It appears that they used it for creating noise rather than inflicting casualties. The Chinese did, however, use another type of "hand grenade." They would fill small earthenware pots with highly poisonous snakes and hurl them at the advancing enemy. When the pots landed and burst, the surprised reptiles were quite agitated, and would promptly sink their deadly fangs into any available body. These "snake bombs" were quite effective in slowing down an advancing enemy—even the most valiant warriors would scatter like children when the things were thrown. One great warlord is reputed to have complained to his enemy that it was disgraceful to use them as they ruined a perfectly good battle.

Sometime later the Chinese replaced the snakes with a mixture of oil and other flammable substances—perhaps the earliest known form of a Molotov cocktail, but still a hand grenade.

Late in the thirteenth century an English alchemist monk named Friar Roger Bacon improved the chemical quality of the old black powder. Sometime around 1304 an Arabian inventor made a bamboo rifle that fired a spear, but he became discouraged when the barrel blew up and left him picking slivers of bamboo out of his body. It was left to another monk, Friar Berthold Schwarz, to develop a relatively usable cannon, and for the next five hundred years cannons and heavy rifles were developed.

There is no mention of grenades until the Revolutionary War; the first accounts of hand grenades being used was by the Continental Marines during attacks on British ships. During the Civil War the Union Army used crude powder grenades, and the Confederate Army retaliated by modifying faulty artillery and mortar shells and turning them into grenades.

During World War II, the United States developed the Mk2 fragmentation hand grenade. During the Korean and Vietnam Wars, the standard grenade became the M26.

The standard hand grenade in use with the Marines now is the M67. This is a fragmentation grenade weighing fourteen ounces. It has a smooth metal body and is shaped like a ball. The inside of the outer case is lined with a tight coil of serrated wire, inside of which is a six-and-a-half-ounce explosive charge known as Composition B. When the pin is pulled and the grenade is thrown, the spring-loaded lever on the side of the grenade flies off and strikes a detonating fuse. This time-delay fuse then detonates the Composition B, which explodes and hurls the fragmented body and wire in all directions.

The M67 grenade has a casualty radius of 15 meters, and an average marine can throw it some 40 meters.

Marine Corps training. After a recruit has fallen on the confidence course, his drill instructor shouts, "Are you a fish, boy?"
"No, Sir!! I am a wet recruit."

In CONGRESS.

The DELEGATES of the UNITED COLONIES of New-Hampshire, Massachusetts Bay, Rhode-Island, Connecticut, New-York, New-Jersey, Pennsylvania, the Counties of New-Castle, Kent, and Sussex on Delaware, Maryland, Virginia, North-Carolina, South-Carolina, and Georgia, to *Samuel Nicholas Esquire*

WE reposing especial Trust and Confidence in your Patriotism, Valour, Conduct and Fidelity, DO by these Presents, constitute and appoint you to be *Captain of Marines* ~~of the Armed called the~~ in the service of the Thirteen United Colonies of North-America, fitted out for the defence of American Liberty, and for repelling every hostile Invasion thereof. You are therefore carefully and diligently to discharge the Duty of *Captain of Marines* by doing and performing all Manner of Things thereunto belonging. And we do strictly charge and require all Officers, Marines and Seamen under your Command, to be obedient to your Orders as *Captain of Marines* And you are to observe and follow such Orders and Directions from Time to Time, as you shall receive from this or a future Congress, of the United Colonies, or Committee of Congress, for that Purpose appointed, or Commander in Chief for the Time being of the Navy of the United Colonies, or any other your superior Officer, according to the Rules and Discipline of War, the Usage of the Sea, and the Instructions herewith given you, in Pursuance of the Trust reposed in you. This Commission to continue in Force untill revoked by this or a future Congress. *Philadelphia Novem. 28, 1775*

By Order of the Congress

John Hancock PRESIDENT.

Attest. *Cha. Thomson Secy*

The commission of Capt. Samuel Nicholas, Esq., USMC. November 1776.

Above: The look of Marines in 1900 in the Canal Zone.

Below: Marines in China in 1933.

A drill instructor tells one of his recruits the difference between wrong and right.

At Parris Island a recruit, pugil stick in hand, takes instruction in the art of self-defense.

A view of new recruits on an early morning run.

The battle for Iwo Jima was one of the Corps' finest hours in World War II.

Left: On Mt. Suribachi, Marines guard the flag they fought so hard to raise.

Marines hit the enemy with a continuous barrage of mortar fire.

Korea: First Division Leathernecks counter fire with fire when attacked by well-entrenched Chinese troops during the division's heroic breakout from Chosin.

A Marine lance corporal walks on point during action in Vietnam.

U.S. Marines proudly raise the American flag on the merchant ship SS *Mayaguez*, after taking possession of the ship which had been captured by the Cambodians.

Four of the five weapons just described are the present basic arms of the Marine infantry rifle squad. The machine gun has, in the past, been more of a platoon weapon, but it is being used more and more at the squad level.

Rifle Squad Deployment

A rifle squad is made up of thirteen men, the squad leader, usually a sergeant, and three "fire teams" of four men each. Each fire team consists of a corporal, who is the fire team leader, a lance corporal, who is the automatic rifleman, and two privates or privates first class. One of the privates is designated as grenadier/rifleman, and the other as rifleman/scout.

The sergeant in charge of the squad will be armed with an M16, which will ordinarily be set to fire in the semiautomatic mode; he will also carry a sidearm. He is responsible to the platoon commander for the discipline, appearance, training, control, conduct, and welfare of the squad at all times—and for the care and condition of its weaponry and equipment. In combat he is also responsible for the fire discipline, fire control, and maneuver of the squad. He must position himself where he can best carry out the orders of the platoon commander, as well as observe and control the squad. His job is command and control; he only fires his weapons in critical situations.

Each corporal will be armed in the same manner as the squad leader, with the exception of the sidearm. The corporal's job is to carry out the orders of the squad leader, and he is responsible to him for the effective employment of his fire team, its fire discipline, and fire control. He is also responsible for his team's weapons and equipment. In carrying out the orders of the squad leader, he must position himself where he can best observe and control the team. Usually he will be close to the automatic rifleman in order to exercise control over the automatic fire as quickly as possible. In addition to his primary duties as a leader, but not to the detriment of them, he must also serve as a rifleman. The senior fire-team leader also serves as assistant to the squad leader and will take control of the squad in his absence.

The lance corporals will have their rifles set on automatic. They carry out the orders of their fire team leaders and act as assistants to them. They are responsible for the effective employment of automatic rifle fire and will take over the fire teams in the absence of their respective corporals.

One of the privates in each team will be the grenadier/rifleman with the M203, 40-mm grenade launcher attached to his rifle. He is responsible for the effective use of the grenade launcher and his rifle, and carries out the orders of his team leader.

The remaining private in each team also follows the orders of his team leader and is responsible for the effective use of his rifle and his function as a scout.

In combat every member of the squad will carry a number of hand grenades and as much spare ammunition as possible, in addition to his own personal fighting equipment, water, and limited rations.

The Marine rifle squad concept, composition, and deployment is the result of many years of development and combat experience. It has been proven that the intelligent, aggressive, and effective employment of the rifle squad in combat is the way to success in battle.

7
BATTLEFIELD LOG:
Bougainville—November, 1943

On November 1, 1943, the Marines landed on the island of Bougainville. Like many other Pacific islands, Bougainville was almost unheard of until they arrived. To most of the Marines who saw action on the island, it was not just another miserable battlefield—it was, to put it mildly, a *filthy, stinking, slimy*, miserable battlefield.

The Third Marine Division landed in Empress Augusta Bay to the almost total surprise of the Japanese. Over 60 percent of the division had been landed with adequate support equipment to hold the beachhead before the enemy could respond. Thereafter, the fighting was bloody and savage, but that was no surprise to the Marines—the Japanese never gave anything up easily.

The military importance of the island lay in its potential as an airbase that would bring American bombers within easy striking distance of the major Japanese base on the island of Rabaul. General MacArthur had seen the significance of the island's location and had insisted that it be taken. After the marines landed, the Imperial Japanese Command appeared to have agreed with MacArthur's judgment. They issued instructions to the defending Sons of Nippon: "Fight to the death for the glorious Emperor. Do *not* surrender." In traditional style, the Marines drove the enemy back and started to build the airbase.

The Japanese retreated to the surrounding jungle and the Marines went in after them; the airbase would be useless with the enemy swarming about in the dense tropical forests within sniping and mortar range.

The veteran Marines had had previous jungle experience, but they were struggling badly in Bougainville since they had never experienced such a deep, ancient, tropical rain forest. It rained every day in November, from noon until dusk—you could almost set your watch by it. The scattered clouds would gather quickly into a great dark mass and the rainfall would start. There was no gentle rain on Bougainville; it always poured down with the force of a powerful shower, stinging any part of a person that was exposed. However, the pelting only occurred on the open trails and in the occasional jungle clearings—the thick canopy formed by the tops of the trees prevented the rain from penetrating directly to the forest floor. Instead the rain was either funneled into miniature waterfalls that would cascade straight to the ground at any available opening, or it would just hit some obstruction and splatter. The rain came from every direction, even the ground, as it bounced off fallen leaves and branches.

The dense overhead canopy had other effects—it blocked out most of the light, it prevented heat from escaping, and it eliminated any possibility of a breeze. The normal tropical temperature, boosted by the heat generated from rotting and decaying vegetation, turned the whole rain-sodden area into a massive sauna. It was dimly lit for the most part, with some places almost devoid of any light, and then there was the smell.

The jungle floor was a steaming, stinking, compost heap, and the only consolation was that consistent exposure to it would eventually lessen the nausea. It was no help to get into a clear area for a few days or even a few hours, because re-exposure was just as bad as the initial contact.

There were slimy, miasmal swamps everywhere, which could swallow squads of Marines without a trace. There were sinewy, whiplike vines that seemed to be alive as they constantly entwined themselves around the limbs of the struggling Marines. It took razor-sharp machetes and knives to cut through them; struggling to get free would invariably make the situation worse.

There was an abundance of wildlife and most of it was unpleasant, even to the battle-hardened Marines. Snakes up to ten feet long were commonplace, and they had a passion for finding a sleeping soldier to curl up alongside of, much to the consternation of the waking man. Mosquitoes came in two sizes only: very large and extra large. They also came in two varieties: pint suckers and quart suckers.

Bats are supposed to be creatures of the night, but the Bougainville jungle was Bat Paradise and they flew around with utter contempt for humans. They were evil-looking creatures whose main purpose appeared to be to collide with Marines, and their preferred landing zone was anywhere on the face and neck. In flight, their wings made a high whirring sound that resembled a falling artillery shell which, of course, caused the Marines to needlessly seek shelter. Since bats are associated with horror stories and evil doings in the night, the flying rodents aroused the ingrained fears of most of the marines.

Even the birds of paradise paid scant regard to what was expected of them. Instead of sitting on their lofty perches displaying their colorful plumage, they would dive screeching and screaming on the unsuspecting Marines.

There were leeches and a variety of other unidentifiable blood-sucking insects, and there were lizards and reptilian creatures that were virtual miniatures from the age of the dinosaur. Occasionally, the languid air was pierced by the tortured, humanlike scream of a creature known to the marines as the "banana cat." Those who had seen the creature said it was half insect-eater and half cat; most of those who had not seen it swore they would kill whatever it was, regardless.

When faced with all these natural "enemies," the Japanese threat sometimes paled a little, but not by much—they were relentless in their attempts to drive the Marines off the island. One veteran Marine of the Bougainville jungle was heard to say, "What we ought to do when the war is over, is to give this island to the Japs and make them live on the damned place—*forever*."

In the rotting, tangled mess of the jungle, the average Japanese soldier had a distinct advantage over the average Marine—his size. The Japanese soldier found it easier to move in the dense undergrowth, and when he had to hack a passageway through the awful mess, he did not have to clear as much or work as hard as his American adversary. To follow the enemy by simply enlarging his trail was utter foolishness and often fatal because booby traps were a combination art and science with the cunning Japanese soldier.

The utter dislike, almost pure hate in some, and the fear of that fetid jungle was expressed by almost everyone who had to operate in it. But there were a few individuals who never complained, who seemed perfectly at ease in a place that all others considered to be totally unbearable. That tiny group of Marines, sometimes only one or two in an eighteen-thousand-man division, were creatures of the jungle. They were not weird in any way, and they were not the much-popularized storybook loners. They simply had an ability which others do not have, an ability which manifested itself when it was most needed—it was called "jungle sense."

One such man was Lieutenant William Kay. He had an incredible talent for scouting and was a master of the art of bushcraft—or junglecraft, as it was in the rain forests of Bougainville. His abilities were such that he was simply referred to as "The Fox," and those who worked with him were glad that he was on their side.

It was The Fox who found the geological freak—a small, steep-sided, rocky ridge that stuck up some four hundred feet from the jungle floor. It looked as though someone had taken a giant slab of rock and had stood it up on its end, then covered it with slime, small trees, and the inevitable tangling, choking vines. From sixty feet away in the dense forest it could not be seen, and only a scout of Lieutenant Kay's experience would have spotted it from the ground.

The Fox immediately reported the discovery to Major Donald Schmuck and he, in turn, sent Captain Steve Cibik and his company to man the ridge. Cibik linked up with The Fox, whom he admired and respected, and followed him through the jungle. The rear Marines in the company trailed out a spool of communications wire for the telephone, and the long column moved slowly toward the base of the ridge.

Dark was falling when The Fox pointed out the base of the ridge to Cibik. It was hardly more than thirty feet away and it was still difficult to see. Cibik was curious about what was on top. The Fox did not know, he had only skirted along part of the base. At that point, the scout left to return to the command post and Cibik was left to lead his men up the ridge. They groped around in the dark and finally found a place where rock was not quite sheer. Using jutting rocks and tangled roots, they clawed their way upward in the dark. At the rear of the column, the sixteen men carrying the two heavy machine guns and ammunition cursed and swore at everything in existence.

After thirty minutes of climbing, they had reached a small plateau and Cibik was informed that the telephone wire had run out. He ordered his men to dig in as he connected the telephone to the end of the wire. Digging in was a joke. Underneath six inches of slimy, root-infested earth was solid rock, but the experienced company still managed to build a credible night defensive position.

Cibik's attempts to raise the command post throughout the night failed. He was persistent because he knew that the Marine artillery had been saturating the area, and he wanted the shooters to know where his company was located before they started firing again.

As dawn approached, he sent a patrol ahead to locate the top of the ridge. When it returned it informed him that the crest was a good two hundred feet higher and that the climbing conditions were difficult. The patrol sergeant also informed him that they had found fifteen Japanese foxholes.

Cibik wasted no time. He ordered his men to move out toward the top. Again the clawing, sweating, and swearing started as everyone struggled to reach the crest of the ridge. When they finally got there, they were just about exhausted. The only good thing about the place was the view. The Japanese appeared to have shared that feeling—an examination of the foxholes showed extensive and recent use.

It was a perfect observation post for the main east–west trail on that side of the island since it was the highest point for miles around.

When they had rested for a few moments, Cibik ordered his men to set up a defensive ring, utilizing the existing foxholes on the top and digging new ones on the sides surrounding them. He suspected that the enemy used the ridge during the day, to spot for their mortars and artillery, and that they returned to the base of the ridge to spend the night in a larger encampment. If his thinking was correct, the Japanese should be heading back up the ridge at any time.

The preparations were almost complete when Cibik's suspicions were confirmed. The marines on the perimeter had heard the Japanese making their way upward, seemingly oblivious to the fact that the ridge now had other residents.

It was ambush time, and the Marines knew it as they settled low in the foxholes. The talkative enemy soldiers walked straight into a hail of fire from the Browning automatic rifles and the fight for the ridge began.

Meanwhile, the Marine artillery at the rear had already started its daily barrage and Cibik was getting worried. He still had no contact with the command post.

As the Japanese recovered, they returned the fire with machine guns and mortars. The first mortar round had only just hit the ridge when The Fox appeared trailing a reel of communications wire. When Cibik had not been heard from at the command post, The Fox was sent out to investigate. He had found a break in the wire and had repaired it before moving on. He then found the end of the wire on the small plateau where the company had spent the night and realized that Cibik had run short before reaching his objective. He had promptly returned to the command post, collected a spare reel, and made his way back to the plateau. When he reached it, he noticed a considerable difference— it was torn to shreds. Having received a full salvo from the Marine artillery, the plateau almost did not exist.

The fight was well underway on the ridge as The Fox made the connection; he speeded up when he heard the familiar sound of the Marine artillery shells coming in overhead. He quickly scrambled up the remaining two hundred feet to Cibik's position, reeling out the vital communications wire as he went.

Cibik did not see The Fox approach the top of the ridge; he was more concerned about the shells that had just landed not more than a hundred yards downslope from his defensive position. Out of the corner of his eye, he saw a movement and he swung around with his rifle. When he saw The Fox with the wire, he quickly put the gun down and grabbed the telephone. Within moments he was in contact with the command post and in control of his own threatening artillery.

The Japanese were not too pleased about losing the ridge. They ranged in their mortars and bombarded it, then they sent wave after wave of troops up in an attempt to take it back. Cibik was well dug in but he was losing men. By late afternoon he had called for replacements and reinforcements.

As night fell the Japanese had failed to gain any ground and they had paid dearly for their attempts. There were bodies of dead Japanese strewn all over the slopes of the ridge.

Although there were no attacks against them that night, few of the Marines slept. From experience they knew that the Japanese were experts at creeping into well-defended positions under cover of darkness, and they were quick to cut the throat of any careless marine. No infiltration attempts were made, however, and as dawn came the mortars and artillery started again.

The first massed attack by the enemy was beaten off shortly after dawn and more of the Emperor's dead adorned the ridge. After that failure the Japanese made some sporadic attempts, but there was no major threat. By midafternoon, the smell of the dead Japanese was so bad that one Marine went down the ridge and buried as many bodies as he could. When he returned from his self-imposed burial detail he was white faced and his clothing reeked with the smell of rotting human flesh, but the sickening smell coming up the ridge had eased considerably.

Throughout that night, Cibik was constantly on the telephone directing the artillery to the base of the ridge where he had estimated that the Japanese were gathering. At dawn the shelling stopped and the Marines waited. They were red eyed and their nerves were ragged, but the enemy was in a worse condition. Cibik's well-directed artillery and mortar fire had torn them apart and those who were left to attack were cut down by the riflemen of his company.

By midafternoon on the third day, the Japanese retreated from the base of the ridge into the hell of the jungle. The battle for the geological freak was over. The Fox had found it and Cibik and his company had defended it and held it—and it became known as "Cibik's Ridge."

8
BATTLEFIELD LOG:
Iwo Jima—February, 1945

There was a steady throb of powerful engines and the sound of rushing water against steel hulls as the amphibious tractors and landing craft steamed in a solid line toward the beach. The mixed smell of engine exhaust fumes and salt spray was not that unpleasant; at least not as unpleasant as some of the smells that would be experienced on the beach that lay ahead.

The beach was black. It was made of soft, energy-sapping, sandlike volcanic ash. It was extremely difficult to walk on under normal circumstances; it was almost impossible when a man was burdened down with a fighting pack that weighed nearly half as much as he did. But he needed that pack and everything in it if he was to have any chance of staying alive on the island of Iwo Jima.

Men would often wonder about staying alive in the moments, seconds, minutes, hours, days, nights, and weeks that were to follow; they would measure their life like that on Iwo Jima, if they were lucky. They would question their own sanity sometimes, when they half hoped that they would get hit, not too badly, but just enough so that the corpsmen would drag them back for evacuation to the safety of the ships offshore. They were not cowards, most had proved that before. They believed in what they were fighting for, they had enlisted in the Marines because of their beliefs, but none of that could stop them from feeling scared. Regardless of rank, battle seasoned or not, similar thoughts would run through the mind of almost every Marine on those black beaches of Iwo Jima.

As the laden assault craft crunched onto the beaches, Japanese mortars and artillery opened up in an almost pathetic response. A few vessels were hit, but for the most part the shells seemed to create nothing more than a barrage of water spouts in the surf.

The cannons on the first amphibious tractors had opened up in reply as they neared the beaches. It appeared that the taking of Iwo might not be too difficult. The Marines knew that there were twenty-three thousand Japanese soldiers defending the island, and that their own assault force comprised over sixty thousand men—an almost three to one advantage. They also knew that for six months and ten days the island had been under heavy aerial bombardment.

It had started when Liberator bombers had first attacked it on August 9, 1944. The recently organized Pacific Ocean Areas Strategic Air Force had given the bombing of the island top priority, and on December 8, 1944, the 21st Bomber Command commenced an incredible seventy-two day bombing marathon. On five separate occasions the navy had assisted the bombers with naval task forces that had shelled the island.

In that period of bombing, prior to the Marines landing on the beaches, Iwo Jima had been the recipient of some of the biggest bombs in the war, and it was subjected to more bombings than any other target in the Pacific. The island was only 758 miles from Tokyo, second to last of the stepping stones to the Japanese mainland. The Japanese knew we were going to try and take it—we had to have it, they had to keep it.

What the Marines did not know was that, prior to their landing on February 19, 1945, the Japanese commander of the island, General Kuribayashi, had issued a document to his troops in order to raise their spirits and unite them. The document was found on the bodies of almost all the enemy dead. Part of it read as follows:

COURAGEOUS BATTLE VOW

Beyond everything we shall dedicate ourselves and all our strength to the defense of this, our island.

We shall take hold of bombs, then charge the enemy tanks and destroy them.

We shall enter into the ranks of our enemy and entirely obliterate them.

With every burst of fire we will, without fail, kill the enemy.

Every man will make it his duty to kill at least ten of the enemy before dying.

Until every one of us has been destroyed we shall harass the enemy by guerrilla tactics . . .

This was heady stuff, and we unwittingly helped to fire up the frenzy, another fact the Marines did not know until the fighting started. It was the bombing. For six months the tiny island had been saturated, and the seventy-two day marathon by the 21st Bomber Command had driven half of the Japanese soldiers on the island crazy—it was fear, mental exhaustion, anticipation, waiting and waiting for the attack, miserable living conditions in underground caves and tunnels, and a host of other torments that burden and break minds before bodies.

As the hundreds of assault craft headed for the beaches, the Japanese waited patiently. They did not open up with ravaging fire on the vulnerable landing craft! Was it just that they did not want to reveal the positions of their gun emplacements? Or was it that in their half-crazed and frenzied state they were afraid that the hated enemy might be scared away? Or that they would be deprived of the opportunity to get their revenge and to repay the enemy for the six months and ten days of abject misery and virtual hell they had suffered?

Such questions may seem somewhat academic, but fighting men, particularly their leaders, must know as much as possible about the enemy, preferably *before* the battle commences.

THE FIGHTING ELITE

What happened after the Marines of the 4th and 5th Divisions landed on the beaches is not academic.

The beach on Iwo Jima had two major terraces running parallel to the shore line. They were about five feet high; the first was less than one hundred yards from the water's edge, the second was about a hundred yards further inland. Beyond the second terrace, at the point where the Marines landed, lay the island's major airfield. Slightly further to the north lay another large airfield, and north of that lay yet a third airfield. That there were three airstrips on this pear-shaped volcanic outcrop that was only two miles long by one mile wide indicated the importance that the Japanese had placed upon it. The southern tip of the island, the pointed end of the pear, contained the highest piece of ground—an extinct volcano called Mount Suribachi.

As Mount Suribachi was the highest point on the island, it was obvious that artillery and mortars would be located there. Despite the bombing, the Japanese had managed to save most of their guns—they had pulled them deep underground during the aerial blitz but they were quickly dragged out when the marines hit the beaches.

Four full assault battalions were on the beaches within forty minutes and fighting hell was well under way. The Marines had not reached the first terrace when the guns on Mount Suribachi, and others in the high ground to the north, found their range on the beachhead.

Machine guns opened up on the marines from holes in the sand dunes, from almost completely demolished pillboxes, from behind the rubble of totally destroyed blockhouses, and even from behind tiny clusters of sand. Some of the enemy just lay out on the open beach and paid the price for it, but not before they had taken their toll on the Marines.

The anti-aircraft guns around the first airfield had their barrels trained down toward the beach and they delivered ravaging fire on the advancing Marines. The Japanese on the beach were fighting in a wild and crazy way—there was no organization or leadership, it was just suicide fighting. Some were killed by their own guns firing from behind, but most had to be killed by the marines with flame throwers, well-aimed grenades, bullets, and bayonets.

There was considerable hand-to-hand fighting as seemingly dead Japanese, some with parts of their anatomy completely blown away, would leap up in a gruesome spectacle and try to kill the nearest Marine. Some of the enemy had buried themselves in the sand with explosive packs or mines strapped to their chests. When the advancing Marines were within several feet of them, they would leap up in a shower of black sand screaming their lungs out. It was a sight that seemed to come straight out of Hades. If these crazed creatures were shot at they would instantly explode, killing all who were close by. Most of them blew themselves up, taking nearby Marines with them.

The first assault waves had landed on the beach at 0902 hours; they had gotten no further than the first terrace by 1100 hours. Tanks had been landed behind them, but they could hardly move in the soft sand and they were easily obliterated by enemy guns. Trucks could not even move; all ammunition and supplies had to be hand carried from the water's edge to the front and, although that was not very far away, moving in the sand under withering fire was miserable and often fatal.

By noon, some of the assault platoons had advanced over the first terrace and had made their way to the edge of the airfield. Here the Japanese were organized in depth as they were now trying to hold the Marines to as small an area as possible. This would enable their artillery to concentrate its fire on a narrow strip and cause the most havoc and destruction.

As the assault troops moved away from the water's edge, the support units—Seabees, engineers, pioneers, medical personnel, radio units—started to land. These groups suffered severe casualties from the concentrated fire of the Japanese, much more so than the front-line troops. As the support groups tried to land, they were met by hostile shelling; every single vessel that approached the beach was hit at least once, many were totally destroyed, and the water's edge became a graveyard for man and machine. That the beach master and his landing officers managed to sort out the incredible mess is remarkable. Somehow, the Marines up front were getting their supplies and moving slowly forward. Slowly, in this case, means about a yard at a time—the Japanese defenses were now consolidating.

By midafternoon, the 23rd Regiment had managed to push its lines to the edge of the airfield. In doing so it was forced to take every defended position by total annihilation. No prisoners were taken, none of the Japanese would surrender. They did not even fall back and regroup, they just stood their ground or charged straight at the Marines toward certain death. They fought ferociously and fanatically, and died doing so.

The toll on the marines was high, so high in fact that the reserve units on board the waiting ships had to be brought ashore at 1630 hours. Getting them ashore was a tremendous effort, despite the fact that we had complete control of the air, and that we were laying down concentrated counterbattery fire. The Japanese had staked almost everything on keeping the Marines on the beaches and they were achieving some success.

Along the two-mile stretch of beach the wreckage and carnage was incredible. The debris was so bad that there were only a few places where troops and supplies could be landed. The broken hulks of almost a hundred landing craft showed the cost of getting the Marines ashore. Jeeps, amphibious tractors, tanks, and half-tracks lay crippled where they had become bogged down in the ominous black sand.

Bulldozers and construction vehicles lay smashed on the roads they were trying to build. The cranes that had been brought ashore to assist with the unloading of supplies were keeled over with their booms sticking out at wild angles.

The beach was littered with packs, rifles, clothing, and gas masks—all had been ripped and shattered by shell fragments. Letters from home, and some that were intended for home, were strewn amongst the wreckage and the bodies.

Death was manifest on that beach. Many a battle-seasoned Marine racing free from the landing craft glanced at it once and then looked straight ahead over the carnage, determined to live, to avenge what he saw then, and what he had seen before.

At 1700 hours all units were ordered to dig in and prepare for a night counterattack. The front line was only five hundred yards from the beach, except at the south of the island where the 5th Division had managed to cut across the narrow neck of the island at the foot of Mount Suribachi, thus isolating it.

A small wedge had been driven into this supposedly impregnable island, but it was a pitiful piece of territory for the price that had been paid. Almost 40 percent of the total forces landed had been lost during the day, and it became the most costly D-day of the Pacific war.

During the night, the expected counterattack did not come; the Japanese had decided not to waste soldiers on ineffectual counterattacks. They would fight from concealed positions to force the marines to try and dig them out.

The second day of the assault, February 20, 1945, saw Marines of the 4th Division in bitter fighting with the enemy. By nightfall they had taken the airfield with the assistance of the tanks that had finally managed to get off the beaches. Again, it was by annihilation of the fanatical enemy.

Apart from the airfield, the front line only moved forward by two hundred yards that day. The Marines and their commanders realized that they were up against a different kind of enemy from what they had encountered on previous campaigns. These soldiers of Nippon were not just fanatical—they were determined, intelligent, well led and well armed, and they had prepared their defenses carefully and properly.

The Japanese had built deep reinforced pillboxes throughout the island. They had been set up in such a manner that they could shoot interlocking bands of fire that would cut down whole companies of Marines.

Well-developed camouflage hid all the enemy installations, and the high ground throughout the entire island was riddled with caves and tunnels. It was a massive honeycomb arrangement with layer upon layer of gun emplacements, dugouts, blockhouses, and observation posts. There was almost no cover from the enemy fire—enemy observation was perfect. Whenever the Marines made a move they were being watched, and at the right moment the Japanese would open fire with mortars and artillery that had already been zeroed in during previous use. The Japanese could smother an area with murderous fire, but as soon as the Marine and naval artillery opened up to counter, or prepare an area, the Japanese would retire deep inside the tunnels until the barrage had ceased. They would emerge when the bombardment was over and simply blast away at the advancing Marines. The situation was to be like that for the next twenty-three days—grim and bloody; every yard of ground had to be fought for.

On the third day, February 21, a few hundred yards were again gained with almost continuous hand-to-hand combat on all parts of the front. The fourth day saw even less ground gained as the Marines moved toward the second airfield. The area between the airfields was described by seasoned Marines as the most complex and heavily defended area they had ever seen. The fighting was severe and vicious, and some areas of the front gained only ten or twenty yards.

The Japanese were now using a new weapon, primarily at night. It was a one-thousand pound rocket, or "buzz bomb." It caused a good deal of consternation until it was identified. The Marines nicknamed it the "floating ash can," or "bubbly wubbly." The first ones fired from deep behind the enemy lines cluttered and cranked across the sky like heavy-laden freight trains and finished up in the sea.

Not only were the engines noisy, they also left a huge trail of fire behind them, and as the first of the rockets soared into the sea they were laughed at. However, when the Japanese dropped their range and the rockets started to fall on the beaches and among the troops, things became more serious. The rockets were not accurate, but on that crowded little island it did not matter. It was discovered later that the Japanese soldiers who launched the things were just as scared of setting them off as the marines were of receiving them.

On the fifth day, February 23, the Marines took Mount Suribachi. Forty men from the 2d Battalion, 28th Marines crawled up to the rim of the mountain crater and raised the American flag. This ceremony was photographed by Technical Sergeant Louis Lowery for *Leatherneck* magazine. Meanwhile, someone hunted around and found a much larger flag and used it for a second flag-raising ceremony. In this ceremony, six men raised the giant flag and they were photographed by Associated Press photographer Joe Rosenthal. This picture first saw light in the photographic dark room on the island of Guam and was to become a symbol of all Americans. The Pulitzer Prize-winning photograph showed five Marines and a Navy corpsman raising the flag. Three of those Marines later died on Iwo Jima, not knowing that they had become very visible in a historic moment.

With the taking of Mount Suribachi, most of the shelling on the beaches stopped and supplies flowed ashore almost unhindered. However, the fighting continued in its slow, brutal, and intensely bitter manner for the next twenty days. It was a deadly game of nerve-racking hide and seek. The Japanese would not surrender, they had to be destroyed wherever they were found. It was not until 1800 hours on March 16, twenty-six days and nine hours after the marines had landed, that the island was declared secure.

Of the twenty-three thousand Japanese on the island, only one hundred and twenty lived. They had never intended to lose the island, but they did, and the greatest battle in Marine Corps history was over.

9

MISSION AND PRESENT STRUCTURE OF THE MARINES

In order to maintain its role as the nation's force in readiness, the Marine Corps must be well organized and trained, have the right weaponry and equipment, and be highly mobile.

To maintain the ability to carry out its mission, the regular establishment of the Marine Corps is officially organized into four major components: headquarters, operating forces, supporting establishment, and the Marine Corps reserve.

The Headquarters Group needs little explanation—that's where all the thinking, plotting, planning, organizing, arranging, overall control, order issuing, and infuriating crystal-ball gazing takes place.

The Supporting Establishment includes the Marine Corps schools, recruit depots, supply installations, barracks, bases, air stations, and all other facilities that train, maintain, and support the actual operating forces.

The Marine Corps Reserve provides a trained force of qualified personnel for active duty in time of war or national emergency.

These three groups are basically in the support role, but that does not mean that they play any less of a part in the Marine Corps. Military history clearly shows that lack of support and organized reserves has been responsible for the failure of many an otherwise superb fighting force.

The Operating Force of the Marine Corps, which is manned by over 60 percent of all Marines on active duty, is the fighting arm of the corps. It is split into three main groups: Fleet Marine Forces, Marine Detachments Afloat, and Security Forces.

The Security Forces provide Marines to serve as security detachments at selected naval shore facilities, and at every United States embassy and consulate in the world. The Marines of the embassy and consulate security guards are hand picked for the duty, and they are all volunteers.

Marine Detachments Afloat provide the Navy with additional gun crews and internal security for the ships. They are also available to the ships' captains for use as limited landing parties and in any other capacity that they deem necessary.

The Fleet Marine Forces are the major combat elements assigned the task of meeting the nation's worldwide responsibilities. In order to do this effectively, the forces are divided into two groups—one group serving with the Atlantic Fleet and the other with the Pacific Fleet.

Fleet Marine Force, Atlantic, operates from Norfolk, Virginia. Fleet Marine Force, Pacific, operates from Camp Smith, Hawaii. They are integral parts of the United States Naval Fleet and are under the full operational control of the individual naval fleet commanders. Only the administrative control and unit and individual training are retained by the commandant of the Marine Corps.

Both the Atlantic and Pacific fleet forces contain ground and air combat forces which make up the air-ground combat team that is cherished by the Marine Corps. The concept of the air-ground team is probably more highly developed and practiced by the Marine Corps than any other fighting force in the world.

Each fleet force has a Marine division and a Marine Aircraft Wing attached. The Marine division is the ground element of the team, containing a balanced combat force of three infantry regiments, an artillery regiment, a tank battalion, and an assault amphibian battalion. There are about 18,000 men in the traditional Marine division, of which approximately 800 are Navy personnel (mostly medical). Each division is organized in such a way that it can conduct the initial amphibious assault and then continue with sustained combat operations ashore.

The Marine Aircraft Wing is the aviation combat team, specifically organized to provide complete air support to the division ground forces. The wing will comprise about 150 helicopters and 200 fixed-wing tactical aircraft, with a manpower establishment of some 14,000 Marine and Navy personnel.

In order to maintain the air-ground team aspect of combat operations, each Fleet Marine Force is organized into three basic marine air-ground task forces. These forces are tailored in size to permit operational flexibility and to allow the deployment of the appropriate amount of strength for a given situation (it would have been ridiculous, for example, to send half the Marine Corps to Grenada when 1000 Marine infantrymen could easily do the job). Regardless of the size of the group, it contains all the elements of the total Marine Fleet Force: command, ground combat, aviation combat, and combat service support (including the Navy-support elements).

The smallest of the Marine air-ground task forces is called the Marine Amphibious Unit. It is commanded by a colonel, and normally comprises a reinforced infantry battalion and a composite air squadron, giving a manpower strength of about 2500 Marines and 300 Navy personnel. The unit can react immediately to a crisis situation and is capable of relatively limited combat operations. When committed ashore, the unit is normally supported by its fleet sea base—a recent example of this was the operation on Grenada. In a situation requiring a larger force, the Marine Amphibious Unit is usually considered to be the leading element.

A larger task force is the Marine Amphibious Brigade. It is commanded by a brigadier general and under his control will be a reinforced infantry regiment and a complete Marine Aircraft Group. Manpower strength will be about 14,000 Marines and 1800 Navy personnel, and the brigade will be capable of conducting amphibious operations of limited scope. When the brigade is landed, it can be supported from its fleet seabase, shore facilities, or a combination of both.

The largest of the task forces is the Marine Amphibious Force. It is commanded by either a major general or a lieutenant general, and is normally built around a Marine division and a Marine Aircraft Wing. Its total strength will range upward from 45,000 Marines and 6300 Navy personnel, and it is capable of conducting a wide range of amphibious assault operations. The force is also capable of sustained operations ashore, and can be tailored for a variety of combat missions in any geographic environment. There are three Marine Amphibious Forces (MAFs) in existence: I MAF is located on the West Coast, II MAF is located on the East Coast, and III MAF is located in the central and western Pacific areas.

Presently, the total Marine Corps strength stands at about 197,300 regular personnel.

10
BATTLEFIELD LOG:
Obong-ni Ridge, Korea—August, 1950

In the southeastern part of Korea lies a small town named Tugok. It is set in a hilly area and, to the Marines from Camp Pendleton, it almost seemed like some areas of Southern California. Just south of Tugok a small road runs east and west between a range of hills. Slightly east of the town the road takes a sweeping turn around one of the larger hills before turning northward. On the south side of the road, directly opposite Tugok, is a range of hills which was known to the marines as Obong-ni Ridge. Both this ridge and the town were being overrun by the North Korean People's Army (NKPA) which was attacking from west to east.

The 5th Marines were moving forward from the east in order to stop the advance of the enemy. As the Marines traveled westward along the road, they had a clear view of Obong-ni Ridge on the left, but could not quite see the town of Tugok on the right—it was obscured by a hill to the east. This was Hill 125, and it was around the base of this hill that the road made a sweeping curve before starting northward. Directly south of Hill 125, on the other side of the bend in the road, was yet another small hill which became known as Observation Hill.

As they moved westward, the Marines had been securing the hills on either side of the road. They had secured

Observation Hill and Hill 125, and had started to advance on Obong-ni Ridge and the town of Tugok itself.

On August 17 the 2d Battalion, 5th Marines, led the assault and they were being supported in the air by the F4 Corsairs of Marine Aircraft Group-33 (MAG-33) and by M-26 Pershing tanks of the 3rd Platoon, Able Company. The Marine armor was led by Second Lieutenant Granville Sweet. He supported the advance by moving his four tanks along the road and around the bend of Hill 125. Once around the bend, he dispersed his tanks off the road as they came under heavy fire from antitank guns and mortars. The tanks replied with their 90-mm main guns and .30-inch machine guns.

One enemy mortar scored a direct hit on the number two tank. Lieutenant Sweet's gunner promptly wiped out the mortar position and then his own tank took a hit from an antitank gun on the lower facing slopes of Obong-ni Ridge. The M-26 rattled as the antitank projectile struck it, but no real damage was done. The number two tank was also undamaged after the mortar strike and it promptly wiped out the antitank gun that had just hit the lieutenant's vehicle. The score was now even.

As the tanks changed positions to break the enemy's aim, the number four tank was shaken fearfully by two antitank projectiles that slammed into its turret. When the shock of the explosion had subsided, the tank commander realized that no damage had been done, and he immediately sought out the offending enemy weapons.

The number two tank had spotted both the guns that had fired the rounds on tank four and immediately silenced one of them with a well placed 90-mm shell. The crew was reloading the 90-mm with a round that was intended for the second enemy gun, but the North Koreans were quicker, and their aim was good. The breech of the number two's gun had barely closed when the enemy's shell struck the turret. Inside the cramped vehicle the crew were jarred and shaken but, again, no real damage was done. As number two was sorting itself out, number four had recovered and it immediately silenced the second of the two offending guns.

Meanwhile, the number three tank, which had so far remained untouched, was engaged in a duel with another antitank gun that was concealed halfway down the northerly slope of Obong-ni Ridge. This particular enemy gun was well dug in and had fired three near misses on its opponent. In reply the tank had fired two rounds, and both of them had looked like hits, but on both occasions the tank crew was startled when another projectile was hurled back at them from the same spot.

The surprised Marine crew was aiming its own third round when the enemy gun let fly with another round, which hit the tank square on the turret. Later, the tank commander admitted he had been so engrossed in the duel that he had forgotten to move his tank after firing the first two rounds. He had obviously given the antitank gunners plenty of time, and practice, at aiming at one spot. It was a fundamental mistake, but he learned from it.

The shuddering was settling down and the tank crew was realizing that they were still in one piece when another two projectiles struck them from two different guns. The fierce rattling and shuddering started again, but no real damage was done, apart from an increase in the level of fright and adrenaline. The driver was immediately ordered to move the massive piece of armor—quickly.

Meanwhile, Lieutenant Sweet's gunner silenced one of the offending guns and tank four took care of the second, but the original gun was still intact. Tank three had no sooner settled in its new position when the enemy gun started the duel again with another near miss. Tank three, it appeared, was its private domain. The return shot from tank three finally ended the duel as the 90-mm shell obliterated the irksome antitank gun. Everyone cheered inside the hot iron belly of tank three and the commander called for another move—he was not going to make the same mistake again.

Tank two, having been hit by one mortar and one antitank projectile, was also on the move. As it neared a shallow depression close to a scraggy-looking tree, it was hit by another antitank projectile with little effect. Once in the depression, it started to bring its gun to bear on the hill again.

It had barely stopped moving when it took two direct hits from enemy mortar. The forty-two-ton tank appeared to be lifted off the ground in a twisting motion, and then dropped.

Inside the tank it was bedlam. Everything that was fastened down seemed to come loose at once and the crew was battered and bruised as they, too, were flung around in the confines. Incredibly, when the dust settled, apart from the mess inside, the armored fighting machine was still perfectly operational. Deciding that this shallow depression was not really a safe place after all, tank two moved again, to the surprise and relief of the watching ground forces and the rest of the tanks. Once out of the depression, her 90-mm barked in anger and another antitank gun fell.

The infantry fought its way up Obong-ni Ridge, but heavy machine gun fire from the top of the hill was slowing the advance. Lieutenant Sweet's tank, along with tank three, had their fire directed to the machine gun positions and successfully wiped them out. But not before both tanks were hit again by several rounds of antitank projectiles. Tanks two and four were also hit, but between them they obliterated the remaining antitank guns on Objective One.

The section then moved further along the road to the west, in the company of the 1st Battalion, which was by now relieving the advance troops of the 2d Battalion. From their most forward position, they blasted several targets in the town of Tugok itself and wiped out a few more antitank guns.

When the infantry had taken the northern tip of Obong-ni Ridge, Lieutenant Sweet recalled his tanks and led them back to the east of Observation Hill for refueling and replenishment of ammunition stocks.

They were a happy bunch of tankmen as they traveled eastward. They had accounted for themselves and they now had more faith in their tanks. All four combined had been hit by a total of twenty-three antitank projectiles, and poor old tank two had taken three direct mortar hits. Neither the tanks nor the crews were bothered too much; only one man was slightly wounded.

As the elated tank crews were being replenished at

their own command post, the gravel crunchers who had taken the northern tip of Obong-ni Ridge noticed four of the enemy's dreaded T-34 tanks in the company of a column of North Korean infantry, moving along the road toward Tugok.

The F4 Corsairs of Marine Aircraft Group 33 were called up and they screamed in on the group. They succeeded in routing the infantry and destroying the last tank in the column. The remaining three tanks trundled eastward relentlessly along the road toward the bend where Lieutenant Sweet's tanks had been so successful. Within seconds of hearing of the approaching T-34s, Lieutenant Sweet had his armor clanking back along the road at its maximum speed of thirty miles per hour.

He had already formulated a plan for taking on the feared T-34s which, until now, had never come face to face with an M-26 Pershing. The T-34 had earned its reputation in smashing the army's M-24 light tank and the M4A3 medium tank—so far, it had defeated all the armor that had been pitted against it.

Lieutenant Sweet and his platoon were very familiar with the M4A3 tank and its 75-mm or 105-mm gun. It had been their own mount until just before they left from the States some six weeks before. Prior to their departure they had been informed that they were to be equipped with the heavier and more modern M-26. The company's commanding officer had been frantic; he did not want his men completely unfamiliar with the new battle tank, particularly the 90-mm gun, on the day they went into action. His persuasiveness with the high command managed to get two of the new machines released for one day. He then took his entire company to the ranges, where each gunner and loader was allowed two rounds. The 90-mm guns were not fired again until the tanks reached Korea.

As they clattered along the road toward the bend beneath Hill 125, Lietuenant Sweet and his men were well content with the M-26. Their old M4A3 machines could not have taken the hammering they had received earlier in the day, and they felt confident that they could take on the T-34s.

Sweet ordered his tanks to load with 90-mm armor-piercing shells. His lead tank, containing one of his best crews, was commanded by Technical Sergeant Cecil Fullerton, and he ordered him to stop short of the bend on the narrow road.

As the M-26s waited for a classic encounter—tank versus tank—the ground pounders on Hill 125 to Sweet's right, and those on Observation Hill to his left, were preparing a reception for the menacing T-34s.

The 1st 75-mm Recoilless Gun Platoon was on Observation Hill and the rocket section of the 1st Battalion's antitank assault platoon was on Hill 125.

Sergeant Fullerton waited patiently on the eastern side of the bend. He knew his adversary was close when he saw the first rocket released by the antitank platoon on Hill 125. He dropped down into his turret and slammed the hatch closed. The 3.5-inch rocket caught the first tank on its left track just as it was about to round the bend. The black hulk responded immediately by firing wildly in the direction of the hill, but continued to move.

The recoilless 75s opened up from the right front of the tank and blasted the turret armor and right track. The monster started to catch fire as it continued to wobble around the bend, still firing its 85-mm main gun and 7.62 machine gun. As it rounded the curve, it came face to face with Fullerton's M-26.

Fullerton's crew released two rounds in smart order—even Fullerton was amazed at the reload and fire speed. Both 90-mm armor-piercing shells slammed squarely into the T-34, which immediately exploded in flames. One North Korean managed to get out of the burning vehicle, but unfortunately for him he was in the middle of a fire fight and was immediately cut down by heavy machine gun fire.

The second T-34 charged toward the bend. Once again, a 3.5-inch rocket from the assault squad found its mark and slammed home. With its right track damaged, it weaved crazily around the curve and was promptly hit in the gas tank by another rocket. It then met the fury of the recoilless 75s and lurched off the road behind the first tank. Its 85-mm gun, however, was still firing wildly into the hills.

The turret hatch of the tank opened and the North Korean commander was trying to get out, when the 1st Battalion rocket man fired a 2.36-inch white phosphorus round. The rocket struck the lid and ricocheted down inside the turret, taking the enemy soldier with it into a blazing inferno. The second tank of Sweet's platoon had squeezed up beside Fullerton's and both M-26s fired in unison. Twelve rounds later, the turret was ripped apart and the hull exploded.

The third T-34 had been a little behind the first two, but it had picked up some speed and it raced around the curve behind the blazing hulks of the first two. It was met with a thundering salvo from the two M-26s, the recoilless rifles, and the rocket platoons. It shuddered to a stop and erupted in a violent explosion within seconds.

Within five flaming minutes the Marines had shattered the myth of the T-34, using not only the M-26 tank, but also every antitank weapon then available to marine infantry. Tankmen and infantry alike were now much more confident in their ability to defeat the enemy armor, and they pressed on with the Battle of Naktong.

One week later the enemy was routed and driven back to the west of the Naktong River. Army units relieved the assault troops of the Marine Corps and occupied the embattled area. The brigade was pulled back from the front for a rest. Its depleted ranks were filled by the arrival of eight hundred marines who had just come from California.

On September 5 the North Koreans launched a massive counterattack in an attempt to retake the ground they had lost. The Second Battle of Naktong had begun; the enemy quickly overran the army units and was back on Obong-ni Ridge and in the town of Tugok.

The Marines were immediately sent back to the front. This time they would be fighting on familiar ground—their objectives were the same as those fought for two weeks earlier.

Hill 125, to the right of the famous bend in the road, was quickly taken, as was Observation Hill to the south. The enemy was resisting fiercely, but the marines were slowly regaining the ground. Once again, just like on August 17, a column of enemy armor started along the road toward the bend by Hill 125. This time there were two T-34s and one tracked armored personnel carrier with the usual line of enemy infantry.

The Marine assault commander, Captain Francis Fenton, Jr., had earlier ordered the tanks of the 1st Platoon to stop on the road to the east of the bend. The weather was bad—fog, mist, and rain. There was no possibility of air support, which was probably the reason the enemy was making a concentrated assault in daylight.

When Fenton first saw the armored column, he reported it to his commander, Lieutenant Colonel George Newton, but at that moment his radio went dead. As luck would have it, every other radio in the company area was inoperative because of the mud and rain. So Fenton could not warn the 1st Platoon's tanks that the enemy was approaching the bend from the west. The rocket firing assault platoons were well aware of the advancing T-34s, however, and Fenton had them scrambling to get into position. The burned-out hulks that still occupied the bend were testimony to the marine infantry skill with antitank weapons.

As the tankmen peered through the fog and rain toward the bend, they could see the charred wrecks of the previous assault. What they could not see, and did not even know existed at that time, were the two enemy tanks about to round the bend. The gun barrels of the M-26s were pointing to their own left front, in the direction of Obong-ni Ridge, as they had been assisting the infantry with covering fire. Consequently, when the first T-34 poked its nose around the bend, the leading M-26 had its 90-mm gun pointing a quarter of a turn away from it.

The lead T-34 tank fired on the Marine vehicle as soon as it came into view. Before the turret of the M-26 could be turned to take aim, several more 85-mm projectiles struck. The brigade then lost its first tank to enemy action. The second M-26 in the Marine column tried to squeeze past to get a shot at the enemy but it, too, was knocked out by 85-mm fire. Both crews of the stricken tanks managed to get out through the escape hatches, with some of the wounded being assisted by members of the Marine engineer mine-clearance team accompanying the tanks.

The road was now blocked, and the remainder of the tank platoon was unable to get at the T-34s. However, the infantry rocketeers had gotten themselves into position and they plastered the lead T-34 with 3.5-inch rockets. The black monster stopped moving and its gun started to fire wildly. Fenton's infantry was then joined by the assault platoon of the 1st Battalion which brought its own 3.5-inch rockets into action.

Within minutes, the infantry had completed the destruction of the first tank, knocked out the second, and destroyed the enemy personnel carrier. When the firing stopped the bend at the foot of Hill 125, as seen through the rain and mist, had become a graveyard of armor. Eight steel brutes were sprawled there in their death—two M-26 Pershing tanks of the 1st Provisional Marine Brigade, one armored personnel carrier, and five of the once-dreaded T-34 tanks of the North Korean People's Army.

It was the Marine infantry that had won this armored battle, without assistance from either air, artillery, or its own tank support. Word spread—gravel crunchers could really knock out T-34s by themselves, and morale amongst the infantry was high. It moved a step higher when an enemy prisoner said the word was out: "Beware these soldiers who wear leggings and always want their dead men back!"—a warrior's tribute to the Marines of the day, who wore knee-high leggings and whose battlefield tradition is to recover their dead and wounded.

11
BATTLEFIELD LOG:
Nui Vu, Vietnam—June, 1966

Dusk was falling as a flight of helicopters settled down momentarily on the slope of the mountain known as Nui Vu. It was twenty-five miles west of Chulai and rose to a height of fifteen hundred feet, dominating the surrounding terrain. To the marines it was known as Hill 488.

Eighteen heavily armed Marines rapidly disembarked from the helicopters and started to make their way up to the top of the hill. This was the 1st Platoon, Charlie Company, 1st Reconnaissance Battalion. As the helicopters pitched up and swung away from the hill, Sergeant Jimmie Howard, the platoon's acting commander, gave a casual wave to one of the pilots, then turned his attention to the task at hand.

Three narrow strips of reasonably level ground ran along the top of the hill for several hundred yards before falling abruptly away down the steep slopes. From the air they vaguely resembled the blades of an aircraft propeller; Howard chose the blade which pointed north for his command post. He placed observation teams on the remaining two blades, giving the platoon a good field of view of the surrounding area.

The top of Nui Vu was an excellent vantage point for Howard's platoon, and he quickly discovered that it had also been used by the enemy. Empty Viet Cong foxholes covered the area; each had a small shelter scooped out some two feet under the surface. Howard allowed his men to use the tiny one-man caves as shelter from the elements and for concealment. There was no other cover or concealment on the top of the hill—just rocks, knee-high grass, and some sparse scrub growth.

Surrounding Nui Vu was an area of steep hills and twisting valleys. It was a veritable bandit's lair and was being used effectively as a training ground by the Viet Cong and North Vietnamese.

Marine intelligence reports showed that the enemy were amassing by the thousands in these hills and valleys. They were planning and practicing for raids against the heavily populated coastal towns and villages. But the enemy leaders were not packing their troops into a few vulnerable assembly points. They were keeping their units widely dispersed, moving mainly in squads and platoons, and would only gather in strength just before an attack.

To lessen the threat, and to keep the enemy off balance, the marines launched Operation KANSAS. Instead of sending infantry batttalions to beat the bushes in search of small enemy groups, they decided to counter with eight- to twenty-man teams of reconnaissance Marines. If the recon teams located a large enemy concentration then the infantry could be flown in. If, as was expected, they came across numerous small groups of enemy, they were to take care of them by calling in air and artillery strikes.

Operation KANSAS was working well with the efficient recon teams. Sergeant Howard's platoon was one of those teams.

For two days after the insertion of his team, Howard was constantly calling for fire missions, as bands of the enemy were seen almost every hour. Some of his requests for air and artillery strikes were denied by his battalion commander, Lieutenant Colonel Arthur Sullivan. The colonel was concerned that the platoon's position would be spotted by a suspicious enemy, despite the fact that Howard would only call for strikes when there was an observation plane playing decoy by circling the area.

After two full days both the colonel and his executive officer, Major Allan Harris, were becoming alarmed at the risks involved in leaving the patrol on the hill for much longer. Despite the fact that the hill was an ideal observation post and Howard had experienced no difficulties, it was decided to leave them on top of Nui Vu for only one more day.

The colonel and his executive officer were justified in their concern—the enemy leaders were well aware of the platoon's presence and were carefully planning an attack.

The reconnaissance teams of Operation KANSAS had been harassing, disrupting, and punishing the Viet Cong and the North Vietnamese in territory the enemy claimed to control completely. The VC/North Vietnamese decided that if they could annihilate Sergeant Howard's unit it would discourage and demoralize other Marine units.

The enemy made their preparations and planned their attack well. On June 15 they moved, completely undetected, a highly trained and well-equipped battalion to the base of Nui Vu. Late in the afternoon, hundreds of enemy troops started up toward the top of the hill with one intent: to wipe out Howard and his seventeen men.

Two members of the Army Special Forces foiled the enemy's surprise attack. The two men had been operating in the vicinity of Nui Vu that same afternoon, and they had watched elements of the North Vietnamese battalion moving toward the hill. Howard had deliberately set his radio on the same frequency as the special forces and he overheard them reporting, to their own headquarters, the enemy's strength and movement toward the hill.

On hearing the report, Sergeant Howard acted quickly and decisively in forming a defensive plan. He gathered his team leaders and briefed them on the situation. He selected a main assembly point and instructed his men to remain in their present positions until the first sign of the approaching enemy, at which time they were to withdraw to the assembly point. The corporals and lance corporals crept back to their individual teams and briefed them in the growing dusk. The 1st Platoon, Charlie Company, then settled down to watch and wait.

Meanwhile, the two Special Forces men tried desperately to persuade the Vietnamese Defense Group, whom they were training at the time, to attack the enemy from the rear. They suddenly found themselves in a nasty situation—their trainees would have nothing to do with the idea and became very hostile.

As Howard waited quietly at the top of the hill, he heard the Special Forces men reporting the situation to their HQ. They were quite vocal in their disgust and anger at not being able to help the Marines, and the language they were using over the radio was certainly not that taught at communications school. Howard did not object; the Special Forces had already helped.

Lance Corporal Ricardo Binns had placed his observation team just forty yards down the slope from Howard's position. They were lying in a shallow depression discussing their sergeant's warning as 2200 hours approached. Binns casually propped himself up on his elbows and pulled his rifle butt into his shoulder. His three men watched as he pointed the barrel at a bush and, without saying a word, fired. The bush pitched backward with a scream and fell thrashing, not twelve feet away. The rest of the team jumped up and each of them threw a grenade, grabbed their rifles, and started to scramble up the slope. Behind them, grenades exploded and automatic weapons rent the air. The battle of Nui Vu—Howard's Hill—had begun.

At the start of the shooting, all of the observation teams immediately withdrew from their outposts to Howard's main position. It was a tiny rock-strewn knoll, not much of a defensive position, but it was the best on the hill. At least the rocks would provide some protection.

Howard placed his two radios behind a large boulder and supervised the positioning of his men in a tight circular perimeter, not more than sixty feet in diameter. Each Marine had a good firing position, and Howard could see each of them from his central position.

When Binns shot the first of the enemy, the North Vietnamese were no more than one hundred and fifty feet from the Marine's main defensive position. As Howard set up his defense, the enemy set up to attack. They had the Marines completely surrounded.

A shrill whistle rent the air and moments later Chinese stick grenades showered the marines from all sides. Some just bounced off the rocks, some rolled back down the slopes, some did not even explode; however, some landed by the Marines and exploded, causing the inevitable casualties. Immediately after the first shower of grenades, the enemy opened up with four .50-caliber machine guns. The guns were positioned around the Marines so that each covered a different quarter of the defensive circle. Their heavy explosive projectiles arched in and shattered, smashed and ricocheted around. Bright red tracer rounds from lighter machine guns streaked toward the Marines, pointing the way for the enemy troops. Sixty-millimeter mortar shells slammed down on the beleaguered Marines, adding large fragments of rock to the metal shrapnel screaming and whining through the air. By this time, ten of the eighteen Marines had been wounded, and one was dead. Then the enemy charged in a well-coordinated attack. They were directed by clacking bamboo sticks and shrill whistles. They threw more grenades, they were yelling and screaming, and firing every weapon they had on automatic.

Staff Sergeant James Earl Howard was in his mid thirties; he was a veteran of Korea and had been wounded three times. He had been awarded the Silver Star for bravery and he was no stranger to what was going on around him now. In the early moments of the battle, he was concerned about his platoon. The troops were young, they had been shocked by the ferocity of the attack, by the screams of their own wounded, and by the situation, which looked hopeless. He need not have worried. The young Marines reacted savagely: the first assault enemy troops only got within twenty feet of the perimeter before they were cut down.

Injured marines were calling for help from almost every position. As the platoon corpsman, Billie Don Holmes, crawled forward to help a wounded man, a grenade exploded in front of him and he lost consciousness. The assault had failed to gain momentum and the enemy at the rear had enough sense not to copy the mistakes of the dead. The enemy now went to the ground and started to probe the perimeter in an attempt to find a weak spot. Small groups tried to crawl close to individual Marines and overwhelm them with a burst of fire and grenades. The Marines, however, were not to be caught off guard. They would listen carefully and then throw a grenade in the direction of any movement they heard. They could throw farther and more accurately than the North Vietnamese and, as the American grenades had double the blast and shrapnel effect of the Chinese grenades, the Marines were successful in forcing the enemy back to regroup.

During the lull, Howard contacted Lieutenant Colonel Sullivan and informed him that his escape route was cut off and that there were just too many of the enemy for them to handle. He requested help to get out. Sullivan called on the Direct Air Support Center of the 1st Marine Division and asked for flare ships, helicopters, and fixed wing aircraft to be sent immediately to Nui Vu. But, for some reason or other, the response was delayed and Howard and his men had to face another massed attack.

It was just past midnight when the enemy struck, again accompanied by screaming, shouting, bamboo clacking, and shrill whistles. The Marines threw the last of their grenades and fired their weapons semi-automatically to conserve ammunition. Now they had to rely on accuracy to stop the charging hordes, and it worked—the enemy fell back once again.

As the North Vietnamese retreated, Howard realized that all the living members of his platoon, including himself, were wounded. He instructed the living to take the ammunition off the dead. He now doubted that they could repel another massed and determined charge by the enemy. Combat experience also told him that the enemy, having been badly mauled twice, would now wait and listen for sounds that would indicate that the Marines had been sufficiently reduced in numbers to be weakened and demoralized. Howard knew what would come next.

As if on cue, high sing-song voices floated up the slopes, screaming taunts and insults that Marines had heard in other wars and other places.

"Marines, you die tonight . . ."

"Marines, you die in one hour . . ."

The psychological battle had now started and Howard knew how to handle it. He instructed his men to shout back anything they wanted. The black, moonless night rang with all the insults and profanity the wounded Marines could think of, but Howard held the best until last. He had briefed his men, and when the next sing-song taunt had died down, he gave the signal. The Marines burst out laughing, as hard and as loud as they could, and they kept laughing for a long time.

The North Vietnamese did not mount another major attack.

As this was going on, Howard could hear the aircraft gathering overhead—attack jets and helicopters. He talked to the pilots but they were waiting for a flare ship to arrive, as they could see nothing on the ground.

At 0100 hours, an Air Force flare ship with the call sign "Smoky Gold" came on station overhead. Howard talked to the pilot of the aircraft and the first flare was dropped. The mountainside was lit up, almost better than daylight.

As the embattled Marines looked down the slopes they were horrified. Some felt it would have been better *not* to have known what was there—hundreds of North Vietnamese reinforcements filling the valleys below. The mountain resembled an ant hill!

The frustrated pilots of the attack jets and armed helicopters did not share the Marines' feelings. They wanted to see. They swarmed in. The first jets concentrated on the valley floor and the approaches of Vui Nu. Rockets hissed down from the bellies of the aircraft and blanketed large areas. The "Huey" helicopters came in behind the jets and scoured the slopes, coming as low as twenty feet, sometimes lower. They raked the mountainside with long, sweeping bursts, then pulled away to spot for the attack jets. The flare ships kept the "lights" on as the jets dropped high explosive and napalm. Then the helicopters came in again to pick off the stragglers, survey the damage, and direct another jet attack.

Two Hueys remained over Howard's position for the rest of the night. When fuel was running low on a Huey, another would be sent out from base to take its place. There were four helicopters assigned to alternate overhead and the enemy tried unsuccessfully to shoot them down, although they did hit two of the four machines during the night.

The helicopter pilots fulfilled dual roles—they strafed the enemy and directed the bomb runs of the jets. With the aid of the flares, the jets could distinguish all the prominent terrain features, but they could not spot Howard's perimeter. To mark specific targets, the Huey pilots instructed "Smoky Gold" to drop flares right on the ground. The jets would then screech in and obliterate the spot.

Howard, by this time, was identifying his position by flicking a refiltered flashlight on and off. The Huey pilots, guided by this, would strafe the area within seventy feet of the Marines. In the shifting light of the wind-blown flares, the pilots were afraid of hitting the Marines, and they had to leave some space in front of Howard's perimeter unexposed to their fire. It was into this space that the enemy crawled, and the fight on the ground continued.

The Marines were now involved in a war of hide and seek. They had run out of grenades and had to rely on cunning and marksmanship. Howard instructed his men only to fire single shots at identified targets. The enemy fired all automatic weapons, they threw stick grenades, and the marines threw back rocks!

But it was a good tactic. A Marine would hear a noise and toss a rock in that general direction. The North Vietnamese would think it was a grenade and dive for another position. Then the Marine would roll or crawl to a spot where he could see the position, and wait. After a few minutes, the North Vietnamese would raise his head to see why the grenade hadn't exploded, and the Marine would fire one round. The range was usually less than thirty feet.

Corpsman Billie Holmes regained consciousness in the light of the flares. He saw a North Vietnamese dragging away the dead Marine beside him and felt another North Vietnamese trying to drag him away. Lance Corporal Victor was lying on his stomach behind a rock. He had been hit twice by grenades since the first flare had gone off and he could hardly move. He saw the enemy dragging away the dead Marine. He took careful aim and fired. The man arched backward and fell dead. He saw the other enemy tugging at Billie Holmes's inert body. He took careful aim again and squeezed the trigger. The bullet hit the man between the eyes and he slumped over Billie Holmes's chest. Victor was amazed when he saw Holmes stir and push the fallen enemy from his chest and then crawl toward him. His reward was to have his own wounds bandaged by Holmes, whose own left arm was lanced with shrapnel, and to get some relief from the terrible pain.

For the rest of the night, Billie Holmes, with his lacerated arm and badly swollen face, crawled from position to position tending the wounded and firing at the enemy.

The bitter skirmish around the perimeter continued, and at 0300 hours, a flight of H34 helicopters whirled over Nui Vu in an attempt to extract Howard and his men. But the intensity of the enemy fire kept them from landing, and Howard was told he would have to fight until dawn.

Shortly afterward, a ricochet struck Howard in the back. His voice over the radio faltered and died. To those listening, the Special Forces, the pilots, and the high-ranking officers of the 1st Marine Division at Chulai, all seemed to have been lost.

For hours they had listened to Howard's voice and, in the hell around him, he had been remarkably cool. Now, the listeners felt personally deprived, particularly the Special Forces. There was relief, even cheering, when Howard's voice came back strongly. He had refused to take morphine from Holmes, fearing the drowsing effect he knew it had. He was unable to use his legs now, and for the remainder of the night he pulled himself from hole to hole encouraging his men and directing their fire. Everywhere he went he dragged a companion—his radio.

Lance Corporal Binns, the man who had fired the opening round, was doing likewise. Despite severe wounds, he crawled around the perimeter. He risked his life again and again gathering enemy weapons and grenades for the Marines to use. He continually encouraged the men and gave assistance wherever needed.

Howard had spoken to the Special Forces camp at Hoi An, some three miles to the south, and had promised to hold his position if they could stop another company of the enemy from climbing the hill. The Special Forces literally whipped their South Vietnamese trainees into doing just that for the besieged Marines.

The night passed and dawn came. At 0525 hours, Howard roared out, "Okay, you people, reveille goes in thirty-five minutes!"

At exactly 0600 hours, his voice rang out again, "Reveille! Reveille!"

It was the beginning of another day—and the perimeter had held.

As the light improved, the Marines had a clear look at the hill. On all sides of their position lay enemy bodies and equipment. The North Vietnamese normally raked the battlefield clean, but such was the accuracy of the Marines' fire that they had left unclaimed almost everything that was close to the perimeter.

Although the firing had slackened off, the badly mauled enemy did not intend to leave. They had the Marines cut off and they seemed intent on destruction, despite the aircraft overhead and the Marine sharpshooters. The North Vietnamese slipped into their holes and prepared to wait for nightfall again.

They continually sniped at any Marine who dared show his head. At least two of the enemy's .50-caliber machine guns were still operating sporadically and so was its mortar.

Help came in the form of the Marine Infantry, Charlie Company of the 5th Marines. They were flown to the southern slope of the mountain by helicopter and they quickly disembarked and started up the hill. They ignored sniper fire and wiped out small pockets of resistance. The mortar team wiped out the North Vietnamese mortar with their first round. One weapons platoon commander cut down a sniper at five hundred yards with a tracer round from his M14.

Charlie Company finally reached Howard's position. They had to climb over enemy bodies on the perimeter to get in and they were amazed at the state of the recon patrol. All the survivors were badly wounded, and they had only eight rounds of ammunition remaining between them.

The fighting on the hill continued until noon. The enemy was tenacious and paid the price—they lost the battle of Hill 488, now known as "Howard's Hill."

Of the eighteen-man reconnaissance team, six were dead. The remaining twelve all received Purple Hearts. Fifteen of the patrol were recommended for the Silver Star. Binns and Holmes were recommended for the Navy Cross. Howard was recommended for the Medal of Honor.

They deserved the recommendations. They had held against incredible odds—eighteen Marines against an estimated eight hundred to one thousand of the enemy. In Howard's own words, "The movie *The Longest Day* was but a twinkle in the eye compared to our night on the hill."

12
WOMEN IN THE MARINES

The successful integration of women in the Marine Corps is a tribute to the Corps itself, and to the men and women of the Corps who worked together for the ideal, not because they had to, or because of acute societal pressures, but because they wanted to. Often their plans were blocked or delayed because of the existing laws of the nation and, along with the individuals and committees who recommended the changes, the Marine Corps waited in frustration for the laws to be amended.

The attitude of women in the Marines is perhaps best expressed by the words of a Marine captain, a mother of two children, who said, "I like my job in the Marines, and when I walk through someplace . . . like Los Angeles International Airport . . . and people stare at me, I'm unashamedly proud that I am one of a special few . . . a member of the United States Marine Corps. I am a Marine!" She then hastily added, "Despite the fact that wearing combat boots is not all that becoming for a lady as they leave an obvious ring around your legs, just above the ankles."

On August 12, 1918, Josephus Daniels, secretary of the navy, authorized the Navy and Marine Corps Reserves to accept women for service. One day later, in Washington, D.C., Opha Mae Johnson became the first of the forerunners of today's women Marines. She was the first of the "reservists (female)," who were enrolled to perform clerical jobs, freeing male clerks for combat duties.

The opening of the ranks of the Corps to women was the result of a government survey which stated that "forty percent of the work at Marine Corps Headquarters could be performed by women as well as men."

Once authorization was given by the secretary of the navy, recruiters were instructed to enlist only women of "excellent character and neat appearance," with business and office experience. The greatest demand was for competent stenographers, bookkeepers, and typists. But women who possessed a working knowledge of correspondence and basic clerical skills were also eligible.

Less than a month after the initial recruiting call, thirty-one women reservists had been signed up, and by September 1918, the commandant of the Marine Corps had called them to active duty. Most were assigned duties in Washington, D.C., at the Marine Corps headquarters, in the offices of the paymaster, quartermaster, adjutant and inspector, and the office of the commandant. A few were stationed in recruiting offices outside the Washington area, some even getting as far away as San Francisco, and Portland, Oregon. (By today's standards that might not seem such a great thing, but in 1918 it was really something to be heralded.)

The situation did not last. The First World War had ended and, on July 30, 1919, Major General Commandant George Barnett issued orders for the separation of all women from the reserve. Those on active duty were immediately transferred to the inactive status and disenrollment gradually continued until 1922, by which time all women reservists had returned to civilian life. Many of the women accepted civil-service appointments at Marine Corps headquarters and retained their contact with the Corps. But from a military standpoint, the Marine Corps remained strictly a man's world for the next two decades; the woman's touch became a matter of memory.

The wounds of Pearl Harbor were still hurting and the Pacific war was almost a year old when the Marine Corps again turned to woman power to meet the incredible demands for personnel. On November 7, 1942, the first wartime commandant of the Marine Corps, General Thomas

Holcomb, approved the formation of the Marine Corps Women's Reserve, under legislation sponsored by Congressman Melvin J. Maas of Minnesota (the late Major General Maas, U.S. Marine Corps Reserve).

Before the official announcement of the formation of the Marine Corps Women's Reserve (MCWR), Ruth Cheney Streeter, a mother of four children, was quietly commissioned a major, USMCWR, and was sworn in as the first director of the Women's Reserve by Frank Knox, secretary of the Navy. She held that post throughout the war and reached the rank of colonel prior to resigning her commission on December 6, 1945.

Streeter was not the first woman to go on active duty with the Marine Corps in World War Two. That honor, and the first commission, went to Anne Lentz, Women's Reserve representative for clothing. She was a civilian clothing expert who had helped the Women's Army Corps and originally came to the Corps on a thirty-day assignment to help design the uniform for the Women's Reserve. As a result, she remained with the Corps in the grade of captain, wearing the uniform she helped to design.

General Holcomb publicly announced the formation of the Reserve on February 13, 1943, and recruiters throughout the nation were immediately swamped with applicants.

As the Marine Corps had no facilities for training women, the navy offered the use of its training schools: Hunter College in New York for enlisted women, and Mount Holyoke College in Massachusetts for officer candidates. The first officer class, comprising seventy-five women, entered Mount Holyoke on March 13, 1943. They were commissioned on May 11 the same year. The first enlisted class of 722 women arrived at Hunter College on March 26, 1943, and graduated on April 25, 1943. It was July 1943, before a newly constructed training complex was opened at Camp Lejeune, North Carolina. It housed both officer and enlisted schools, as well as the Women's Reserve specialist schools.

The recruiting slogan was, "Free a Marine to Fight!" and that is exactly what the women did. Within one year women reservists were serving at every major post and station in the continental United States.

It was originally anticipated that there were thirty specific job functions for women to fill. After less than a year, this had grown to more than two hundred. Apart from the usual clerical jobs, specialist billets for which women had been trained were turned over to them. In considerable numbers they were assigned to such fields as communications, quartermaster, post exchange, motor transport, food services, personnel, intelligence, education, legal assistance, and photography. In aviation units, their skills ranged from parachute rigger to control tower operator.

By June of 1944, women reservists constituted 85 percent of enlisted personnel on duty at Marine Corps headquarters, and 50 to 60 percent of the staff manning all major posts and stations in the United States. In September of the same year, regulations were modified to permit women to serve on a volunteer basis anywhere in the Western Hemisphere, including Alaska and Hawaii. Before the war was over, almost one thousand women served with Marine Garrison forces, Pearl Harbor, and at Marine Corps air stations in Hawaii.

At peak strength during the war, the Women's Reserve numbered some 19,000 (division strength), and the second wartime commandant, General Alexander Vandegrift, remarked that women reservists could "feel responsible for putting the Sixth Marine Division in the field; for without women filling jobs throughout the Marine Corps, there would not have been sufficient men available to form that division."

When the war ended, the women's units were rapidly demobilized and, by the end of December 1945, over 65 percent of the Women's Reserve had been transferred to inactive status. Those who remained were scheduled to be released by September 1946.

In August 1946, with total demobilization of the Women's Reserve imminent, the Marine Corps decided that never again did they want to start from scratch. The Corps elected to retain a small group of trained women to set up a postwar reserve, and a few selected women reservists were assigned to Marine Corps headquarters to work out plans for the postwar reserve. A few other women reservists were also retained on active duty assignments at major posts, stations, and recruiting districts.

Between 1946 and 1948, there were no more than a hundred volunteer women reservists on the active duty list. Few though they were, however, their continuity of service bridged the gap between the wartime reserve and the war and peacetime component of the Corps.

With the introduction of Public Law 625 in 1948—the Women's Armed Services Integration Act—the situation changed for the better. The law, which authorized the acceptance of women into the regular component of the Marine Corps, but not to combat-related areas, was a major step forward.

Initially, appointment or enlistment was limited to women then on active duty or with previous honorable reserve service, but in January 1949, the Marines began recruiting women without prior military service. The number of women allowed in the regular component was restricted—it was not to exceed two percent of the total service strength. Women could not hold permanent rank above lieutenant colonel, they could not be assigned duty in aircraft or naval vessels engaged in combat missions, nor to vessels of the navy other than transport and hospital ships. In addition, the director of the Women Marines was to be selected from women officers serving in the rank of major or above and would hold the temporary rank of colonel.

Major Julia Hamblet was director of the Women's Reserve at the time the law was passed and it was on her recommendation that Colonel Katherine Towle, the second director of the wartime Women's Reserve, was appointed director of the Women Marines. Colonel Towle, then serving as assistant dean of women at the University of California at Berkeley, was recalled to active duty. On November 3, 1948, she was discharged as a colonel from the Marine Corps Reserve and accepted a regular commission as a permanent lieutenant colonel in the United States Marine

Corps. On the following day, she was appointed director of women marines, with the temporary rank of colonel. Colonel Towle and Major Hamblet were among the first women officers to receive a regular commission.

Basic training for enlisted women, which had ended at Camp Lejeune in 1945, was reactivated early in 1949 at the Marine Corps Recruit Depot, Parris Island, South Carolina. The newly formed unit became the 3rd Recruit Training Battalion under the command of Captain Margaret Henderson who, ten years later, would be appointed director of women marines with the rank of colonel. Colonel Henderson, recalling her experiences, wrote, "I readily observed how having women train at a station where male recruits trained resulted in increased responsiveness by the women Marines to the requirements of the Marine Corps." The key to the success of the new program, she noted, was "the guidance, cooperation, and acceptance of the women's training program by the commanding general, Major General Alfred Noble, and his entire staff at Parris Island."

In June 1949, the women's officer's training class was established at Marine Corps Schools, Quantico, Virginia, with Captain Elsie Hill (later to become lieutenant colonel) as the unit's first commanding officer. The commandant of the Marine Corps Schools at that time was Major General Lemuel Shepherd, Jr. (later to become commandant of the Marine Corps) who personally welcomed the first class and assured them of his support. "There is, " he said, "a definite place for women Marines during peace, as there was during war." It was during this period that the Women's Reserve was also being built and strengthened in conjunction with the men's organized reserve units.

When the Korean War erupted in 1950, the Marine Corps Reserve was called up, and it was the first time in history that the Women's Reserve was mobilized. They were ready.

As the Korean conflict drew to a close, Colonel Towle retired and Julia Hamblet once again became the director of women marines, this time as a colonel.

Following the Korean War, the Women's Reserve units grew in strength until 1958, when budget constraints forced their disbandment. About 35 percent of the women who were attached to the deactivated units became affiliated with the men's organized reserve units, their parent organizations. Many of the women in the remaining units who were transferred to the inactive reserve status continued their Marine Corps service as active participants in the reserve volunteer training units.

By 1964, the number of women Marines on active duty in the regular component had dropped as low as 1,448. However, their selection and training continued to provide the Corps with a source of well-trained, professionally qualified women.

As the Vietnam War approached, more and more formal career training programs opened up to women officers. At the same time, advanced technical training programs were being opened up to enlisted women. With the Vietnam commitment came an increase in strength in the women Marines; it peaked at about 2,700 by the middle of the conflict.

At the outbreak of the Vietnam War, there were about 60 women Marines serving outside the continental United States. By 1970, that number had increased to more than 200. Women Marines were on duty in England, France, Germany, Italy, Belgium, Japan, Okinawa, Hong Kong, Panama, and the Dominican Republic. It was in the Dominican Republic, in April 1965, that Master Sergeant Josephine Gerbers Davis had the dubious honor of being the first woman Marine to experience hostile fire, when the American Embassy was attacked.

In September 1966, the commandant of the Marine Corps directed that women Marines be permitted to volunteer for service in the Far East. Within a month, Captain Marilyn Wallace was on her way to the Marine Corps Air Station at Iwakuni, Japan.

On March 18, 1967, Master Sergeant Barbara Dulinsky, who had volunteered for service in Vietnam, reported to the Military Assistance Command in Saigon. She was the first woman Marine ordered into a combat zone.

On November 8 the same year, Public Law 90-130 was

passed. The bill increased, and to a certain extent equalized, promotion opportunities for women in the military. It repealed the legal limitations on the number of women in the armed services, permitted the permanent promotion of women to colonel, and provided for the temporary appointment of a woman to rear admiral or brigadier general while serving in a specified "flag rank" billet in the Navy or the Marine Corps.

Women officers were still precluded from competing directly with male Marines for promotion. But, beginning in 1974, they were selected for promotion by the same board membership as male Marine officers, and a woman officer was added to the selection board. Regulations were also changed to allow the permanent appointment of enlisted women as first sergeant and sergeant major in the Corps.

During the Vietnam era, Colonel Barbara Bishop and, later, Colonel Jeanette Sustad served as directors of the Women Marines. During Colonel Sustad's tenure regulations that had mandated the separation of women who were pregnant or had custody of minor children were modified to permit waivers to remain on active duty on a case-by-case basis.

The Vietnam War ended with the signing of the Paris Peace Accords on January 27, 1973. The draft was replaced with an all-volunteer force, and on February 1 that same year, Colonel Margaret Brewer became the seventh, and final, director of the Women Marines.

In the post-Vietnam era, the Marine Corps took positive steps to integrate women more fully. A specially formed ad hoc committee studied the increased effectiveness and utilization of women in the Marine Corps and, in November 1973, the committee's recommendations were approved by the commandant with the written comment, "Let's move out!"

Among the most significant recommendations were the establishment of a pilot program to train women for duty with selected, stateside elements of the Fleet Marine Forces, the assignment of women Marines to all occupational fields except the combat arms, and the elimination of the regulation that prohibited women from commanding units other than women's units.

In 1974 the commandant approved a change in policy permitting the assignment of women to specified rear echelon elements of the Fleet Marine Forces, on the condition that women Marines not be deployed with assault units, or units likely to become engaged in combat.

In 1975 the Marine Corps approved the assignment of women Marines to all occupational fields except the four designated as the combat arms—infantry, artillery, armor, and pilot/air crew. Some assignment restrictions remained, including the preservation of a rotation base for male Marines, the need for adequate facilities for women, the availability of nondeployable billets for women, and the legal restrictions prohibiting the assignment of women Marines to combat ships and aircraft.

Integration of women into the Marine Corps had been so successful that on June 30, 1977, the office of the director of the Women Marines was dissolved after thirty-four years, and its functions were transferred to other Marine Corps's staff agencies. Colonel Margaret Brewer was appointed the Marine Corps's deputy director of information, the first woman to hold this billet. On May 11, 1978, she was appointed to a general officer's billet as director of information (later changed to director of public affairs) with the rank of brigadier general. In this capacity, she served as the Marine Corps's first woman general officer until her retirement on June 30, 1980.

Finally, on September 15, 1981, the Defense Officers Personnel Management Act became law, and among its many provisions it specified an integrated promotion system for men and women officers. For the first time, women became eligible for selection to various limited duty officer categories. However, the Marine Corps's forward thinking had already seen the value and effectiveness of the women Marines, and was well ahead of the legislature in its integration policies. The principal reason was that the Marine Corps identifies itself as a team with the constitutional commitment "to provide for the common defense." The Marine

Corps had long since reasoned that all members of the team must perform their duties with that concept in mind and, too, that all members must be treated equally within the bounds of function, practicality, honor, and dignity.

13
FUTURE DEVELOPMENT OF THE MARINE CORPS

The circumstances in which the Marines will be used in the future will most likely differ significantly from those that have prevailed before the Korean War. During the thirty years since that war, the strategic supremacy of the United States has been replaced by U.S. and USSR parity. The strength of the Soviet military forces has increased dramatically during the past three decades—particularly the strength of naval forces—while both the image and real capabilities of our own forces have been diminished.

The continuing Soviet perpetration of a multifront, creeping global offensive only increases the possibility of direct confrontation between the U.S. and Soviet conventional forces. The Soviet's larger and more sophisticated navy is a vital link in the support of its territory-grabbing and agitation programs, and the projection of its influence in the third world. Soviet naval forces are now displaying the ability to project their amphibious power far beyond their own ports.

The international economic condition has seen some equally dramatic changes, and the growing dependence of the industrial nations on third world raw material resources has increased their vulnerability to economic blackmail by third world cartels. The chronic political instability of most of the third world further jeopardizes the steady flow of raw materials and has a profound effect on the economics of the

West. Also, the perception of many people in recent years that the United States is unable to deter or to react effectively to international terrorism has further aggravated the situation.

All these realities must be kept clearly in mind when it comes to building and developing our own military forces. The United States is a nation dependent upon unobstructed sea lanes for the free flow of raw materials and other trade. American, Western European, and Japanese interdependence of markets and products is critical to the economic and political positions of all concerned and, in most cases, dependence upon raw material imports will continue to grow.

The same interdependence has turned NATO into a multi-ocean alliance with global responsibilities. The United States, as the strongest among the industrial nations, has the major responsibility for the preservation of the common interest. That responsibility requires our presence in the principal oceanic theaters, along with a credible power projection capability. Such theaters include Central America, South America, the Pacific Basin, Northeast, Southeast and Southwest Asia, the Indian Ocean Basin, the Persian Gulf, and Southern Africa.

Historically, the value of timely action with credible forces tailored for the situation is clearly evident, the Grenada incident being the most recent example of this. History has also shown that promised "friendly" bases and the right to overfly "friendly" nations quickly disappears when a crisis starts to deepen. It is also interesting that in the past twenty-five years there have been approximately 250 crises to which the United States has responded by deploying forces. Naval forces have been used in over 200 of the events, and amphibious forces in over 80 percent of the cases. This frequent naval involvement results directly from their state of readiness and their mobility.

In order to meet the changing and ever-present threat of armed conflict against forces with modern and sophisticated weaponry, the Marine Corps must constantly upgrade and improve its arsenal of weapons. It must also change and adapt its tactics base of good intelligence and well-founded principles. The Marine Corps has never been noted for lagging behind in anything, and the Corps's planners and strategists are tireless in their efforts to ensure that Marine tactics can meet the threats against the nation's interests in the future.

Weaponry and equipment must be upgraded—improved tactics alone will not defend against an enemy's superior fire power—and to this end, the Marine Corps is presently introducing and attempting to procure new weapons at all levels.

For the ground combat elements alone, and the rifleman in particular, several new weapons have entered service, and one of the first of these is a new, lightweight machine gun for the squad fire team. The M249, 5.5mm machine gun is an individually portable, gas-operated, magazine- or belt-fed machine gun that will replace the M16 as the automatic rifle in the fire team. The M249 has considerably more range and sustained firepower than the M16 in the automatic mode, and it will provide the Marine infantry with firepower equal to that of known and anticipated threat forces.

Another weapon that is entirely new to the Marine Corps is the MK-19 MOD3, 40-mm Grenade Machine Gun. It is a crew-operated, belt-fed automatic grenade launcher, capable of engaging light armored vehicles and infantry from 65 meters to 2200 meters with a high-explosive, dual-purpose round capable of penetrating two-and-a-half inches of rolled homogeneous armor. The maximum effective range of the MK-19 against point targets is 1600 meters. It is being introduced to counter the growing numbers of infantry fighting vehicles with which potential enemies are equipping their forces, and it will allow the antitank weapons to concentrate their fire on the tanks, without worrying about the enemy's infantry and light armor.

Yet another new weapon to the Marine Corps is the Assault Rocket Launcher. It is a man-portable assault weapon to be employed at the rifle company level. It is capable of defeating field fortifications (bunkers), urban targets (concrete and masonry), and has a secondary capability of destroying light armor. It employs a dual-mode warhead which automatically discriminates between relatively soft targets (such as earth and sandbags), and hard targets (concrete, masonry, and armor). The warhead functions in the delay mode against the soft targets and in the immediate detonation mode against the hard targets. This weapon is a necessary and valuable addition to the rifle company as the Marine Corps currently does not possess a man-portable assault weapon capable of defeating field fortifications and urban targets.

The infantryman's defense against enemy fixed wing and helicopter attack has become a considerable problem in modern times, and the threat is increasing. To combat this, the Marine Corps has procured the Stinger Missile. It is a man-portable, visually aimed, shoulder-fired, surface-to-air defense weapon system designed to counter the low altitude aircraft threat. The weapon can engage jet and helicopter aircraft from all aspects, including head on. By employing a passive, infra-red homing missile, precision intercepts are assured forward of the hot jet plume of a high-speed aircraft. The missile is equipped with an advanced infra-red countermeasure system. In order to prevent the possibilities of shooting down a friendly aircraft, the Stinger also incorporates an IFF (Identification, Friend or Foe) interrogator, which prevents the operator from shooting down one of his own aircraft.

Light armored fighting vehicles are being introduced in ever-increasing numbers by threat forces. The Marines have evaluated the system and now have their own light armored fighting vehicle. It is an eight foot by eight foot wheeled vehicle weighing fourteen-and-a-half tons, and is amphibious and helicopter transportable. The basic vehicle will add a new dimension to the force commander's tactical employment concept—he will have a fully integrated combined arms unit possessing significant firepower and tactical mobility.

The Marine Corps is currently looking at a replacement for its aging main battle tank, the M60A1, and the likely replacement will be the M1E1 with a 120-mm stabilized cannon and improved armor. This new tank will be a modified version of the army's M1 tank.

There are many new pieces of equipment that the Marine Corps is presently employing. They are far too numerous to detail here, but they include: new and improved tracked landing vehicles (LVT), medium range surface-to-air missiles, laser range-finding equipment, advanced anti-armor missiles, pistols, mobile projected gun systems, communication and navigation equipment, night vision equipment, and air cushion vehicles.

As this is being written, some piece of equipment in the Marine Corps arsenal is becoming obsolete. It is a mutation process, either you will adapt or you will die. The Marine Corps has to adapt, but it must do it carefully, as mistakes can be costly, both in terms of life and money. Unlimited funds are just not available for indiscriminate military purchasing in our free nation, and the Marine Corps has a reputation for judicious and prudent spending.

The Marines have always believed that they will be the first to fight in any conflict that requires combat, and they are continually training and preparing for it.

It is hard to believe the state of readiness that the Marine Corps continues to maintain; this requires good equipment, patience, training, discipline, pride, leadership, and teamwork.

The Marine Corps has set high standards for itself and for the nation. When they are not actually engaged in some hostile conflict to uphold the interests, dignity, and freedom of the nation and its allies, they are practicing for it ... practice, practice, practice—it never seems to stop.

The Marine Corps is the nation's force in readiness for one simple reason: they are, without doubt, warriors who are always ready.

14
BATTLEFIELD LOG:
Grenada—October, 1983

In the third week of October, 1983, men of the 22d Marine Amphibious Unit boarded their ships in Morehead City, North Carolina. Their destination was Beirut, Lebanon, where they were to assist in the peace-keeping mission already underway.

As the ships steamed out of the port, they headed east toward the Mediterranean, but not for long. Out of sight of the shoreline they slowly turned southward. Experienced Marines and sailors noticed almost immediately the southerly heading, but expected it to change back to the east at any time.

It is not unusual for warships to head off in a direction other than the posted one. This could be for a variety of reasons: to rendezvous with escorts, to aid a stricken vessel, to avoid bad weather over the horizon or, as sometimes is the case, to engage in some form of clandestine, en route exercise with unseen submarines or aircraft. This sort of exercise requires only the presence of the vessel or vessels in a certain area or on a particular course, it does not require everyone knowing what for, or why. Quite often, even the ship's captain has no knowledge of the reason. He obeys orders and directs his ship this way and that, in accordance with instructions given.

One famous naval captain actually knew what his destination was, but couldn't figure out how he was going to get there from the preliminary sailing instructions he was given. He was heard to remark to an inquiring sailor, "Just because the ship is heading in that direction it doesn't mean that we're actually going that way." The befuddled sailor relayed the comment to his buddies and one of them immediately snorted, "He's finally flipped! He's been on this old rust bucket too long. They should make him an admiral now."

At first, the old salts of the 22d paid little attention to the southerly direction, but as the course held, hour after hour, they became suspicious. If they had asked their commander, Colonel James Faulkner, he could only have told them that their orders were to head south to a certain point and then head east to the Mediterranean. He only knew one thing more than they did; he knew the point where the turn eastward would be made. But the turn to the east never came.

Colonel Faulkner received his orders: *Make an amphibious landing in Grenada. Help protect and evacuate Americans and designated foreign nationals. Maintain peace and order so that the government of Grenada would have sufficient time to take control of the country.*

The word spread like wildfire throughout the ships. Combat! It would be the largest U.S. combat action since the war in Vietnam!

"Where the hell's Grenada?"

That question was asked again and again, and hundreds of Marines and sailors went scouring about in kit bags and lockers for maps. There was no shame in asking the question; many an educated man would be as much in the dark as the average Marine when the tiny island's name hit the headlines.

When they finally gathered together some maps and information on their target, Grenada was seen to be the southernmost island in the East Indies, not more than ninety miles off the coast of Venezuela.

The island was approximately 133 square miles in size and was populated by about one hundred thousand people. It was first settled in the seventeenth century by French colonists and became a British possession in 1783. Grenada, one of the smallest nations in the world, received full independence with the British Commonwealth of Nations on February 4, 1974.

For a number of years increasingly disturbing events on the island had the small surrounding nations, as well as the United States, concerned. When the Prime Minister of Grenada, Maurice Bishop, was assassinated and his Marxist revolutionary council was deposed, the neighboring island nations requested help from the United States. The 2d Fleet was steaming into Caribbean waters in response to that call for help.

The Marine Amphibious Unit had linked up with elements of the 2d Fleet, whose flagship was the giant aircraft carrier, the USS *Guam*. On board the great carrier was the military commander of the Grenada Task Force, Vice Admiral Joseph Metcalf. It was on his shoulders that the responsibility for the complete operation would rest.

The staff of the 22d Marine Amphibious Unit was given forty-eight hours to plan the landing operation, with most of the work load given to Major Ernie Vanhuss. For the next few days he and his staff were to get almost no sleep.

Operational planning began on board the USS *Guam*, the Amphibious Unit's headquarters at sea, and it quickly became apparent that very little intelligence information was available on either the island's terrain or on the opposing forces.

The Marine staff were not the only ones to have sleepless nights. The ground forces of the 2d Battalion, 8th Marines, were also kept busy. Their combat gear had to be prepared and packed, their weapons cleaned and checked again and again. Live ammunition was readied for issue, but would only be given to the troops just prior to disembarkation. Landing craft and amphtracs (amphibious tractors) were readied. Tank crews scurried around preparing the iron monsters, loading their cramped interiors with ammunition and supplies. Everything that needed fuel was topped off, one of the most dangerous tasks in the confines of crowded ships. The huge assault helicopters took on an even more awesome appearance as crew chiefs secured their machine guns and loaded lockers with ammunition.

The wicked-looking Cobra attack helicopters seemed to devour incredible amounts of rockets, ammunition, and fuel. When readied, they sat on their hangar decks looking like fat, contented dragonflies.

The Marine stores and supply group personnel were run ragged—everybody wanted everything at once. Besides issuing all that was needed for the initial assault, they had to inventory and prepare the equipment to support the logistical end of the operation once the troops were on the beaches.

Mechanics and technicians never knew they were so popular. Almost every driver or commander of anything mechanical seemed to want something checked or fixed. As one veteran gunnery sergeant said, "There's nothing like word of an upcoming fight to make those lazy bastards want to shape up their equipment." But he, more than most, knew the real reason: nerves.

Teamwork was also in evidence throughout the 2d Fleet—they would supply additional firepower, transportation, supplies, and shelter.

During the hours of darkness on the night of the 24th, the fleet approached Grenada. The first objective of the Marines was to take and secure Pearls Airport in the northeast section of the island. The landing craft were to approach the beaches at 0500, H-hour, but the Navy SEAL's reconnaissance units reported that there was dense coral and that the surf was running some six feet high.

Colonel Faulkner changed the plan of attack—it was now to be a heliborne assault. The landing craft personnel relaxed a little and the helicopter crews became a little more nervous. So did some of the veteran Marines, who preferred a beach assault with tanks and armored vehicle support right alongside them, or at least right behind them.

As H-hour neared, twenty-two heavily laden helicopters lifted off the deck of the carrier *Guam* and headed toward Grenada. All were flying without lights and, as they approached the island, the weather got worse. The two spearhead helicopters were the fast Cobras. They arrived over Pearls Airport in darkness and circled the target area for five minutes in an attempt to draw fire as the eighteen giant troop-laden helicopters and two more gunships approached. Three 12.7-mm guns on the hill to the north of the field swallowed the bait and opened fire. They were shooting so erratically that the pilot of the lead Cobra reasoned they must not have been able to see the helicopters in the darkness and bad weather.

With a target now identified, the Cobras attacked. They were supported by the remaining two gunships, which had just arrived with the troop carriers.

The troop helicopters dropped quickly into the landing zone, and Echo Company Marines disembarked in rapid order and dispersed to secure the area. They were surprised at how bad the weather was, but it did at least give them some protection from enemy fire. The troop carriers scuttled back to the USS *Guam* and were quickly refueled and reloaded with Marines; this time it was Fox Company. They lifted off and headed back toward Pearls Airport in a tight group.

Fox Company was to be placed at Landing Zone Oreo, which was just to the south of Pearls. It was not going to be an easy task as the zone could only take one helicopter at a time. As the first helicopter landed and the Marines raced out, enemy mortar fire started. That got everybody's attention, particularly the helicopter pilots', who then ceased in their attempts to make perfect landings. They flew quickly over the landing zone and unceremoniously dropped the huge machines smartly to the ground. The helicopter crews didn't have to do much bidding to the battle-clad Marines cramped inside. Some of the crew chiefs were almost trampled on as the troops tumbled out and dispersed, but they didn't complain; they, too, were eager to get away from the landing zone.

The small landing zone under fire brought back memories to some of the veteran Marines. The first sergeant of Fox Company briefly described the journey from the ship to the landing zone: "You have a tendency to age fast. I felt the helo hit the ground and then bounce up again. I closed my eyes . . . another hot LZ!"

The four hundred men of Echo and Fox companies started sweeping and clearing the area around Pearls. For the most part, they found that the enemy would start out quite bravely but would then give up quickly after a few minutes of the Marines' concentrated fire. Occasional pockets of strong resistance were encountered, but the disciplined and coordinated attacks of the infantry and the gunships soon overran them. The airport area was declared secure by 0731 hours, and both companies started to move out toward the island's capital city, St. George's.

Close daylight reconnaissance of the northeast coast only confirmed what the SEALs had discovered the previous night—the surf and sea conditions were still not suitable for landing craft to approach. The tank-landing ships *Manitowoc* and *Barnstable County* were ordered to steam with all haste around the top of the island to the west coast.

It was 1930 hours as the landing ships and their escorts approached the beach known as Grand Mal. The SEALs had given the all clear and both vessels steamed in steadily, grounded their bows on the coral sand, and dropped their landing ramps. Five tank and thirteen amphtrac drivers gunned their engines and clattered down the ramps onto the beach. An assortment of jeeps and 250 Marines of Golf Company accompanied the armored vehicles as they moved up from the water's edge onto the island's tiny roads. The characteristic growling sound of the powerful main battle tank engines, and the high whine of the amphtrac engines faded from the beaches as the armored column moved inland toward the town of Fort Frederick. The objective was to drive out a strong force of Cubans who were surrounding the governor's residence, and then proceed to Fort Frederick.

During the early part of the afternoon, Fox and Echo companies joined together just outside St. George's and secured a landing zone at the local racetrack. They established a solid perimeter defense and Colonel Faulkner flew in and set up his command post. No attempt was made to enter the capital—that would come later.

With steel tracks clattering and clacking, the column traveled slowly throughout the night, stopping only to clear out small pockets of resistance. As they neared the governor's residence, the fighting became quite fierce, but the

Marine infantry with the tanks and amphtracs were well practiced at working as a team. The enemy fire was snuffed in smart order and the Cuban troops were routed. The area was declared secure at 0712 hours, October 26, and the armored column moved on toward Fort Frederick.

The governor, Sir Paul Scoon, an officer of Queen Elizabeth II and her representative to the island, was rescued along with his staff. He was immediately flown out to the USS *Guam* to be the guest of Vice Admiral Metcalf and the United States Navy.

At 1600 hours, a combined Marine and Ranger helicopter assault was launched against the Grand Anse School area. Cuban soldiers put up a hail of fire as six Marine helicopters evacuated the American medical students to the carrier *Guam*.

Throughout the afternoon the tireless Fox and Echo companies were on the move again in the sector north of the capital, clearing out pockets of resistance with the help of the ever-present gunships.

Meanwhile, Golf Company and the armored column were heading toward Fort Frederick and fighting a running battle against well-trained Cuban and some Grenadian soldiers. As the column approached the fort, they came up against heavy fire from 75-mm recoilless rifles and 12.7-mm guns. Once again the teamwork of armor and infantry, this time assisted by air support, made short work suppressing the enemy. Fort Frederick was secured at 1725 hours.

The Marines discovered that this was the island's strong point: it housed the total command and control system for the People's Revolutionary Army. Teams of specialist intelligence officers were flown in with extra troops to hold the fort. Fuel and ammunition were air-lifted in for the tanks and amphtracs, and Golf Company had some hot food that night.

As dawn was breaking on the morning of the 27th, Golf Company and the armor were on the move again. By 1150 hours they had secured Fort Lucas after another spasmodic running battle. Thirty minutes later they moved out; the running battle started again as they sped toward Richmond Hill Prison. They met increased opposition at the prison but it did not last long, even though the Marines had to be careful with their fire for fear of injuring civilian prisoners.

With the area secure, they moved on again and by 1700 hours they had reached the Ross Point Hotel. After a brief fire fight, the hotel was taken and seven civilians were rescued.

It had been a productive day for Golf Company and it now dug in for the night.

During the same day, Fox Company saw plenty of action on the way to Fort Ruperd. It had some troublesome moments with several snipers as it neared the fort, and had to hold up the advance until its own sharpshooters took care of the situation.

The opposition at Fort Ruperd was the worst it had encountered. The enemy were well hidden in the bushes and trees surrounding the area and it took some effort to dig them out. Mortar and heavy-caliber machine-gun fire were used somewhat erratically against them, but it was effective enough to make everyone take more care and slow things down. Finally, Fort Ruperd was taken and the Marines captured such a vast amount of enemy weapons and ammunition that the company commander ordered a squad to stay behind to guard the small arsenal until helicopters could be brought in to remove it.

Echo Company was not inactive. It was sent into the hills to the north of Pearls Airport and instructed to conduct a sweep into the island's interior from there. Some fierce fighting ensued in the hills as the Marines fought and captured several 12.7-mm gun emplacements. As they drove the enemy back, they discovered the reason for the resistance: a small arsenal of Soviet-made weapons and ammunition. The company called in a helicopter and had the weaponry and munitions taken out. Then they continued their spearhead operation into the interior. Once again they met with stiff resistance, and only hard fighting and concentrated fire drove the enemy back.

Soldiers of the 82d Airborne linked forces with Golf Company on the morning of the 28th, and the combined units moved with the armored column toward St. George's. The Marine units spearheaded the attack and encountered some heavy fighting during the initial stages. Again, discipline and training showed as the Marines and the airborne units carefully eliminated the opposition. By 1455 hours, the capital was declared secure.

By this time, the Marines and Army forces on the island had a firm grip on the situation and the remaining enemy opposition was dwindling, but it was not over.

Early in the afternoon on the 29th, a Marine patrol was attacked just to the south of Pearls and the startled and angry Marines retaliated and destroyed their attackers in less than twenty minutes.

Late on that same afternoon, Golf Company was on the move again. It raced up the coast to patrol the townships of Victoria and Gouyave to help maintain order as rowdy elements and straggling bands of the remaining enemy were trying to intimidate the townspeople. With the arrival of Golf Company, the trouble ceased.

Tactical operations for the Marine Amphibious Unit were winding down and stability and order was returning to Grenada. The attitude of the people of Grenada toward the Marines was one of overwhelming gratitude. By the end of their stay, the tired and dusty Marines were more than satisfied with their work on the island because of the reaction of the Grenadians.

On the night of October 31, all the Marines were back on board their ships. The island was left in good hands with the 82d Airborne and elements of a police force from neighboring islands.

The Marines left behind them an impressive record. The infantry combat element had overcome some stubborn opposition and had captured large numbers of Cubans and Grenadian left-wing extremists. They had also taken huge stockpiles of enemy weapons and ammunition and, by the end of the swift operation, they had driven through and walked over almost the whole of the northern part of the island. They had been particularly careful to avoid casualties among the island's mostly peaceful population, despite the attempts of some of the enemy to use the civilians as cover.

But the Marines were still not yet quite finished. The

small Grenadian island of Carriacou was reported to be a Cuban stronghold and at 0430 hours on November 1, the landing craft, amphtracs, and Marines hit the beaches again. No resistance was encountered on tiny Carriacou and it was declared secure. The following morning, at 1000 hours, elements of the 82d Airborne landed on Carriacou and the Marines returned to their ships.

Shortly afterward, the 22d Marine Amphibious Unit set sail for the Mediterranean—their course this time was easterly, without doubt.

One of the items they had removed from the island went with them—it was a joke for which the Grenadians no longer had use:

> *Question:* Why do Grenadians go to Trinidad for their dental work?
> *Answer:* Because it's the only place they can safely open their mouths.

ABOUT THE AUTHOR

IAN PADDEN was born and educated in England. During service with the British Military he learned to fly and also developed an interest in specialized reconnaissance, espionage, and counter insurgency warfare. His interests in these subjects required him to have a thorough knowledge of other special ("elite") military units throughout the world.

He was taught deep-sea diving by Royal Navy instructors and worked as a commercial diver in construction, salvage, and offshore oil drilling. He spent further time in the oil industry working as a driller, drilling supervisor, and drilling engineer and was later employed by one of the world's leading subsea drilling equipment manufacturers as a specialist engineer and training instructor. He left the company to become a drilling consultant, and in that capacity has been responsible for drilling oil wells, both on land and offshore, throughout the world.

One of Ian's hobbies is aerobatic competition flying. He has been a member of the British Aerobatic Team since 1978 and has represented Great Britain in two world championships.

Ian Padden began writing in 1963 when he presented a special paper on "The Foundation, Formation and Operating Principles of the Roman Army" to the British Army School of Education. In 1965 he assisted in the writing of "The Principles of Diving" by Mark Terrell (Stanley Paul: London). During his career in the oil industry, he was commissioned to write training manuals and narrations for training films. He has also written two television scripts and various treatments for documentaries. He is currently finishing a full-length novel.

Join the Allies on the Road to Victory
BANTAM WAR BOOKS

☐	24702	THE FIGHTING ELITE: U.S. MARINES Ian Padden	$2.95
☐	24703	THE FIGHTING ELITE: U.S. RANGERS Ian Padden	$2.95
☐	24810	RUSSIAN SUBMARINES IN ARCTIC WATERS I. Kolyshkin	$3.50
☐	24487	STUKA PILOT Hans V. Rudel*	$3.50
☐	24164	INVASION—THEY'RE COMING! Paul Carel	$3.95
☐	23843	AS EAGLES SCREAM Donald Burgett	$2.95
☐	24656	FLY FOR YOUR LIFE L. Forrester	$3.95
☐	25194	THOUSAND MILE WAR B. Garfield	$3.95
☐	22832	D DAY: THE SIXTH OF JUNE, 1944 D. Howarth	$2.95
☐	22703	LONDON CALLING NORTH POLE H. J. Giskes	$2.50
☐	20749	BREAKOUT J. Potter*	$2.50
☐	25032	COMPANY COMMANDER C. MacDonald	$3.95
☐	24104	A SENSE OF HONOR J. Webb	$3.50
☐	23820	WAR AS I KNEW IT Patton, Jr.	$3.95

*Cannot be sold to Canadian Residents.
Prices and availability subject to change without notice.

Buy them at your local bookstore or use this handy coupon for ordering:

Bantam Books, Inc., Dept. WO, 414 East Golf Road, Des Plaines, Ill. 60016

Please send me the books I have checked above. I am enclosing $_____
(please add $1.25 to cover postage and handling). Send check or money order
—no cash or C.O.D.'s please.

Mr/Mrs/Miss _____

Address _____

City _____ State/Zip _____

WO—4/85

Please allow four to six weeks for delivery. This offer expires 10/85.

Join the Allies on the Road to Victory
BANTAM WAR BOOKS

☐	23718	THREE CAME HOME Keith	$2.95
☐	24127	NIGHT FIGHTER* J. R. D. Braham	$2.95
☐	24163	BRAZEN CHARIOTS Robert Crisp	$2.95
☐	24179	ROCKET FIGHTER Mano Ziegler	$2.95
☐	24264	QUEEN OF THE FLAT-TOPS Stanley Johnson	$2.95
☐	24372	TYPHOON OF STEEL: THE BATTLE OF OKINAWA J. Belote & Wm. Belote	$3.95
☐	23985	SERENADE TO THE BIG BIRD Bert Stiles	$2.95
☐	14516	CLEAR THE BRIDGE R. O'Kane	$3.95
☐	23882	PIGBOATS Theodore Roscoe	$3.95
☐	23890	IWO JIMA R. Newcomb	$3.50
☐	22897	BATTLE FOR GUADALCANAL Samuel A. Griffith, II	$3.50
☐	24624	COMBAT COMMAND F. Sherman	$3.50
☐	23055	WITH THE OLD BREED E. B. Sledge	$2.95
☐	24482	HORRIDO! Ramond F. Toliver & Trevor J. Constable	$3.95

*Cannot be sold to Canadian Residents.

Prices and availability subject to change without notice.

Buy them at your local bookstore or use this handy coupon for ordering:

Bantam Books, Inc., Dept. WW2, 414 East Golf Road, Des Plaines, Ill. 60016

Please send me the books I have checked above. I am enclosing $_____
(please add $1.25 to cover postage and handling). Send check or money order
—no cash or C.O.D.'s please.

Mr/Mrs/Miss_____

Address_____

City_____ State/Zip_____

WW2—2/85

Please allow four to six weeks for delivery. This offer expires 8/85.

SPECIAL MONEY SAVING OFFER

Now you can have an up-to-date listing of Bantam's hundreds of titles plus take advantage of our unique and exciting bonus book offer. A special offer which gives you the opportunity to purchase a Bantam book for only 50¢. Here's how!

By ordering any five books at the regular price per order, you can also choose any other single book listed (up to a $4.95 value) for just 50¢. Some restrictions do apply, but for further details why not send for Bantam's listing of titles today!

Just send us your name and address plus 50¢ to defray the postage and handling costs.

BANTAM BOOKS, INC.
Dept. FC, 414 East Golf Road, Des Plaines, Ill 60016

Mr./Mrs./Miss/Ms. _____
(please print)

Address _____

City_____ State_____ Zip_____

FC—3/84